Applied Mineralogy: Applications in Industry

Applied Mineralogy: Applications in Industry

Edited by Trinity Collins

New York

Published by Syrawood Publishing House,
750 Third Avenue, 9th Floor,
New York, NY 10017, USA
www.syrawoodpublishinghouse.com

Applied Mineralogy: Applications in Industry
Edited by Trinity Collins

International Standard Book Number: 978-1-64740-382-9 (Hardback)

Cataloging-in-publication Data

Applied mineralogy : applications in industry / edited by Trinity Collins.
p. cm.
Includes bibliographical references and index.
ISBN 978-1-64740-382-9
1. Mineralogy. 2. Minerals. I. Collins, Trinity.
QE363.2 .M56 2023
549--dc23

TABLE OF CONTENTS

PREFACE

Minerals are nonrenewable natural resources that are essential in the energy, manufacturing and construction industry. Mineralogy refers to a discipline of science that examines all facets of minerals, including their occurrence and distribution in nature, physical characteristics, internal crystal structure and chemical composition, as well as the physicochemical conditions that led to their origin. Mineralogy has significant role in mineral extraction and advancements in this field have resulted in industrialization of minerals. Applied mineralogy is a research field which focuses on the manufacturing of materials, and the prospection, extraction and refinery of ores. It also studies the impact of minerals on human health and the environment. This book unravels the recent studies in the field of applied mineralogy. It explores all the important applications of this field in the present day scenario. The book is appropriate for students seeking detailed information in this area as well as for experts.

This book is a comprehensive compilation of works of different researchers from varied parts of the world. It includes valuable experiences of the researchers with the sole objective of providing the readers (learners) with a proper knowledge of the concerned field. This book will be beneficial in evoking inspiration and enhancing the knowledge of the interested readers.

In the end, I would like to extend my heartiest thanks to the authors who worked with great determination on their chapters. I also appreciate the publisher's support in the course of the book. I would also like to deeply acknowledge my family who stood by me as a source of inspiration during the project.

Editor

Geometallurgy, Technological Mineralogy and Processing of Mineral Raw

1

Crystal-Chemical and Technological Features of the KMA Natural Magnetites

T. Gzogyan(✉) and S. Gzogyan

Belgorod National Research University, Belgorod, Russia
mehanobr1@yandex.ru

Abstract. The results of researches of crystal-chemical features of magnetites obtained from the KMA ferruginous quartzites deposits are presented. It is shown that all magnetites are represented by cation-scarce mineral differences. The correlation dependences between the technological indexes (yield, extraction) and the magnetite concentration in quartzites are established.

Keywords: Crystal-chemical features · Ferruginous quartzites · Magnetite · Maghemite · Mossbauer spectra · Magnetic properties · Defectiveness

1 Introduction

The most important geochemical feature of iron is the presence of several oxidation degrees. By crystal-chemical properties of the Fe^{2+} ion is close to the Mg^{2+} and Ca^{2+} ions and to the other main elements. Due to crystal-chemical similarity Fe replaces Mg and partially Ca in many silicates. In quartzites, the main ore mineral is magnetite, the crystal-chemical structure characteristics of which determine the technological properties of quartzites, according to which quartzites are characterized by high variability of composition and properties, and the results of enrichment depend on their material composition. It is worth noting the diversity of mineral composition, texture and structural features of ores and, as a consequence, a wide range of physical, mechanical and technological properties (Gsogyan 2010).

2 Methods and Approaches

In this work the crystal-chemical features of the main ore minerals of unoxidized and oxidized quartzite KMA deposits are investigated using modern research methods (nuclear gamma-resonance spectroscopy (Mossbauer effect), X-ray analysis and X-ray fluorescence). Measurements of the dependence of magnetization on the applied magnetic field and saturation magnetization by the vibration method are carried out.

Magnetite is a typomorphic mineral composition and properties of which depends on the conditions of formation and has got a structural spinel. This is the ohFd3 m space group, where cations occupy 16 octahedral (B) and 8 tetrahedral (A) positions. Therefore, for ideal crystals of $Fe_1^{3+}[Fe_1^{3+}F_1^{2+}]O_4$ the reversed spinel, the ratio of the intensities of the leftmost lines of the mossbauer absorption spectra, taking into account the difference in the probabilities of the effect for the octahedral and tetrahedral sublattices, should be slightly less than two.

3 Results and Discussion

The characteristic lines of the obtained Mössbauer absorption spectra of quartzites represent the superposition of a series of six-peak spectra. The ratio of the intensities corresponding to half of the ions in the A and B positions is practically not satisfied, despite the fact that the selected sample is closest to the monomineral difference. In addition, there is also a slight broadening of the leftmost and rightmost lines of the total absorption.

Mossbauer spectra indicate that the intensity ratio 1:2 is not satisfied. Violation of this ratio may be due either to an isomorphic substitution of iron ions in the B-position of $Fe_1^{3+}\left[Fe^{3+}_{5/3-2/3X}F^{2+}_{X-Y}Me^{2+}_{y\square 1/3-1/3X}\right]O_4$ or to the phenomenon of non-stoichiometry and the presence of $Fe_1{}^{2+}$ vacancies in the same position $Fe_1^{3+}\left[Fe^{3+}_{5/3-2/3X}F^{2+}_{X-Y}Me^{2+}_{y\square 1/3-1/3X}\right]O_4$. Obviously, at x = 1, this is the stoichiometric magnetite $Fe_1^{3+}[Fe_1^{3+}F_1^{2+}]O_4$, and at $x = 0 - \gamma - Fe_2O_3\left(Fe_1^{3+}\left[Fe^{3+}_{5/3\square 2/3X}\right]O_4\right)$. The appearance of isomorphically replacing ions in the octahedral position of the magnetite spinel structure leads to a partial removal of electron exchange in position B between Fe^{2+} and Fe^{3+} and, as a result, to a change in the ratio of the intensities of the lines corresponding to iron ions in positions A and B with simultaneous broadening of not only the extreme right groups of lines, but also the entire set of lines of the total resonance absorption spectrum (broadening is especially pronounced with an increase in the concentration of isomorphically substituting ions). The appearance of a cationic vacancy in position B of the spinel structure does not lead to experimentally observed broadening but only to a redistribution of the intensities of the lines corresponding to Fe^{3+} and [Fe^{3+}, Fe^{2+}] in positions A and B due to increasing in the contribution of Fe^{3+} which does not participate in electronic exchange Fe^{2+} with position B due to vacancies. But in both cases the integral intensity remains constant. However, the presence of a cationic vacancy in the octahedral position in the spinel structure significantly affects the technological properties. The combination of experimental data obtained allows us to write the crystal-chemical formula of magnetites in the form $Fe_1^{3+}\left[Fe^{3+}_{5/3-2/3X}Fe^{2+}_{x\square 1/3-1/3X}\right]O_4$, where $\square$ – is a cation vacancy (in the structure of magnetite as isomorphic substituting ions can be Mg^{+2}, Mn^{+2}, Ni^{+2}, Ti^{+2}, etc.), and x – is non-stoichiometric (the defectiveness is the Fe^{2+} deficiency widely varying in area of deposits). At x = 0 we have a six-peak absorption spectrum corresponding to maghemite. The condition of electro-neutrality is observed due to the formation of defects that is, electron holes. The relationship between Fe^{3+} and [Fe^{3+}, Fe^{2+}] determines the behavior of such magnetites in magnetic fields.

In addition, martite lines are present on the Mossbauer spectra of magnetites and this indicates that the crystal lattice of magnetites is coherently connected with the lattice of the "parent" magnetite. The presence of a coherently coupled

$Fe^0_1\left[Fe^{3+}_{5/3\square 1/3}\right]O_4$ structure significantly affects the degree of its extraction during the enrichment. Traces of hematite and various non-metallic particles are present with magnetite.

Thus, magnetite is a very sensitive indicator of the conditions of deposits formation and its "life", started in nature, is manifested and sometimes significantly in the kinetics of technology, because rapid processes intensively destroy its information structure and slowly transfer it to the products resulting from processing (Gsogyan 2010).

Features of heterogeneity of the composition and properties of magnetite affect the variability of its oxidation degree in the course of crushing process and, as a result, the technological indicators of enrichment (Gsogyan 2010). Quantitative XRF of quartzites revealed the presence in them along with magnetite, hematite and maghemite; moreover, the presence of maghemite is the characteristic of all mineral types of quartzite. XRF showed that the highest content of maghemite ($\sim$11%) is the characteristic of magnetite quartzite, somewhat lower ($\sim$8%) is observed in semi-oxidized and the lowest ($\sim$4.5%) in oxidized.

Morphological series observed in nature in ferruginous quartzites: Fe_4O_4 (wustite) → $Fe^{2+}Fe_2{}^{3+}O_4$ (magnetite) → γFe_2O_3 (maghemite) → αFe_2O_3 (hematite) → $Fe_2O_3 \cdot H_2O$ (goethite) → $\gamma FeO(OH)$ (lepidocrocite) → $Fe_2O_3 \cdot xH_2O$ (limonite /martite) emphasizes the complex relationships between their structures. With the transition from one structural type to another, some features of the original type are inherited. The edge of the unit cell, depending on the composition of the oxidation of wustite to magnetite and maghemite consistently decreases with the number of Fe^{2+} cations. Linear dependence of the lattice parameter on the composition emphasizes the similarity between the structures of these compounds. Reducing edges of the unit cell of magnetite in the transition to maghemite caused by the replacement of Fe^{2+} with an ionic radius of 0,80 Å on Fe^{3+} with an ionic radius of 0,67 Å with simultaneous removal of one third of the Fe^{2+} ions from the structure of magnetite. The magnetites associated with the incomplete processes of martitization are highly heterogeneous, which is associated with the manifestation of morphotropic transformations in the series indicated above, the content of Fe in which falls from 72.4% (magnetite) to 62.0% (limonite).

The average edge size of the unit cell varies from 8.371 Å (Lebedinsky) to 8.395 Å (Mikhailovsky) and this is somewhat less than that of a magnetite of stoichiometric composition (8.396 Å). The evaluation of magnetite deformations allows us to conclude that a change in the main interplanar distances of magnetite samples is more characteristic of Lebedinsky and Prioskolsky.

Microprobe analysis revealed iron deficiency in magnetite grains of various generations, the content of which varies from 66.89% (Mikhailovsky) to 72.3% (Stoilensky).

Thermomagnetic analysis in almost all magnetites at T = 340–350 °C shows a weak exothermic effect corresponding to maghemite and its presence is confirmed by electron diffraction patterns, which indicate the absence of periodicity in the arrangement of some reflexes and the presence of defects and dislocations in the magnetite lattice (Fig. 1).

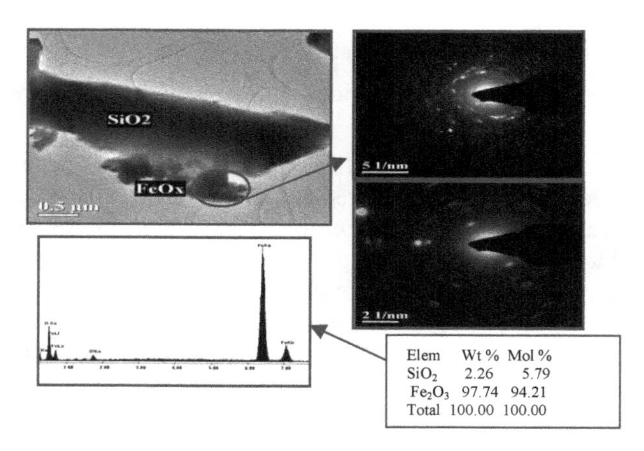

Fig. 1. The electron phase, taken from the magnetite particles at different angles of orientation of the device

Maghemite is marked from 11% (Mikhailovsky and Prioskolsky) to 6% (Korobkovsky). The data of high-temperature magnetometry confirm the lack of Fe^{+2} in the KMA magnetites and their difference from stoichiometric: the Curie point for magnetite averaged 571.8 Å (Stoilensky) to 578.3 Å (Prioskolsky). The change in the ratio of Fe^{3+} and Fe^{2+} causes the inextensive magnetic properties which are one of the most important factors affecting the enrichment. These studies suggest that, due to the lack of a sufficient amount of isomorphic impurities, we are dealing with alternative behavior in the two-phase system of $Fe_2O_3 \div Fe_3O_4$, that is, the formation of metastable transitional phases with a structure closer to the matrix than the equilibrium phase. This is confirmed by the results of determining the concentrations of isomorphically substituting impurities of Mg^{2+}, Mn^{2+}, Ti^{2+} and Ni^{2+}, which did not exceed 0.2% NiO; 0.25% MnO_2 and 0.4% MgO.

All Mossbauer absorption spectra contain maghemite lines. Consequently, the formation of solid solutions of $FeO{:}Fe_2O_3$ coherently coupled with γ-Fe_2O_3 and semicoherently with Fe_2O_3 is strictly observed in deposits. This led to the widespread development of quartzites heterogeneity in their composition and properties. The coherent phases formed in this way can be clusters of dissolved atoms in the lattice of the parent material that is often observed. The presence of "non-stoichiometric" cation-deficient differences of magnetites led to the need to find the value that determines their defectiveness. The crystal-chemical formula of such magnetites is: $Fe_1^{3+}\left[Fe^{3+}_{5/3-2/3X}Fe^{2+}_{x}Fe^{2+}_{1/3-1/3X}\right]O_4$ where x is defectiveness (Fe^{2+} deficiency in the octahedral position of the spinel structure) (for $x = 1$ is ideal magnetite $Fe^{3+}\left[Fe_1^{3+}Fe^{2+}\right]O_4$ (content of 72.4%), and at $x \rightarrow 0$ a complete transition up to γ-Fe_2O_3 (crystallographic symmetry) $Fe_1^{3+}\left[Fe^{3+}_{5/3}\,Fe^{2+}_{1/3}\right]O_4$ when a part of iron ions is recharged to the state of Fe^{3+} and the iron content drops to 69.9%).

The dependence of the magnetic properties on the composition and crystal-chemical features of magnetites is performed by the vibration method in a uniform magnetic field H up to 71.62 kA/m in the temperature range from 20 to 800 °C. The obtained dependence of the specific saturation magnetization has an inflection point which apparently is associated with the influence of a sufficiently large weight fraction of $Fe_1^{3+}\left[Fe_{5/3}^{3+}\ Fe_{1/3}^{2+}\right]O_4$ The unevenness of magnetic properties is due to the change in the ratio of Fe^{3+} and Fe^{2+}. Specific magnetic susceptibility indicates a different petrographic composition of ores and the size of magnetite impregnation and their significant heterogeneity which is a consequence of their genesis and is expressed in the difference in their technological properties. The saturation magnetization is a multiparameter function for ferruginous quartzites.

In this regard, the dependences of technological indicators of enrichment were obtained as a function of the reduced value (σ_s/σ_s^{id}), where σ_s are the obtained values of the specific magnetization of this sample; σ_s^{id} is the specific magnetization of the isolated magnetite fraction (Fe_{tot} = 72.2%, SiO_2 less than 0.2%).

4 Conclusions

- magnetites from ferruginous quartzite deposits of the KMA are represented by cation-deficient mineral differences with a wide variation of the magnitude of defects (x) in the area of occurrence;
- the presence of coherently bound γ-Fe_2O_3 and semi-coherently bound α-Fe_2O_3 is determined depending on the degree of oxidation which undergoes spinodal and intracrystalline decomposition up to hematite;
- technological indicators of enrichment depends on the crystal-chemical and magnetic properties of quartzites. These allow to find optimal conditions for concentrates obtaining, i.e. implement the “geological” management of the production.

Reference

Gzogyan TN (2010) To the question of the heterogeneity of the KMA magnetite deposits. Min Inf-Anal Bull (5):256–259

2

Mineralogical and Technological Features of Tin Minerals at Pravourmiysky Deposit

T. Chikisheva[1,2,3(✉)], S. Prokopyev[1,2,3], E. Kolesov[4], V. Kilin[4], A. Karpova[1,3], E. Prokopyev[1,2], and V. Tukuser[1,3]

[1] LCC PC «Spirit», Irkutsk, Russia
chikishevatatyana@mail.ru
[2] Institute of the Earth Crust SB RAS, Irkutsk, Russia
[3] Irkutsk State University, Irkutsk, Russia
[4] PJSC «RUSOLOVO», Moscow, Russia

Abstract. The paper presents the data obtained in the process of mineralogical studies of technological samples of tin ore from the Pravourmiysky deposit. The authors studied in detail the mineralogical features of ores and tin-containing minerals and their significance for the enrichment technology. During of the study, the information on the main technological properties of the ore of the deposit was clarified and supplemented. As a result of the study of the mineral composition, the mineralogical features of the ore were identified, allowing to select the methods of ore enrichment and predict the quality of the concentrates and products obtained. The causes of loss of tin with tailings were established. The obtained data on the mineral composition, properties of ore minerals, and textural and structural features of the ores will be applied when modernizing their enrichment technology at the processing plant.

Keywords: Mineralogical research · Tin ores · Tin minerals

1 Introduction

In the 21st century, tin is very in demand in the global economy due to its use in new industries, the introduction of innovative technologies and the environmental friendliness of metal. At the same time, the tin market is volatile, depending on sharp fluctuations in the price situation and annual changes between supply and demand [1].

Russian industry consumes about 6.5–7 thousand tons of tin per year. About 90% of the mined tin is imported. Tin deposits in Russia are among the richest in the world. Pravourmiyskoe tin deposit is one of the promising ore deposits of tin. In addition, together with tin, tungsten can be mined from ore deposits. To achieve a more complete extraction of valuable components from the ore, a detailed study of the mineral composition of ores, the textural and structural features of the ore, the physical properties of minerals and the degree of their contrast is necessary. Mineralogical analysis is the basis for the study of the material composition, structure and texture, choice of directions and methods of preparing raw materials for processing, enrichment technologies and metallurgy [2]. the paper is devoted to the study of the mineralogical and technological features of the Pravourmiysky deposit tin ores in the Far East.

2 Methods and Approaches

The object of the research was technological samples of tin ore from the Pravourmiisky deposit. The mineral composition of the ore and the quantitative assessment of the contents of each mineral were determined using the methods of optical microscopy and using x-ray methods. Mineralogical analysis of ore crushed to a particle size of less than 2 mm was carried out according to the methods of the Scientific Council on methods of mineralogical research. The sequence of operations to determine the mineral composition of the ore samples consisted in dividing the initial material of the samples into size classes, followed by gravitational fractionation and studying the distribution of minerals into fractions for each size class.

3 Results and Discussion

The main ore mineral of tin at the Pravourmiysky deposit is cassiterite. Stannin, mawsonite, stannoidite are present in small quantities. More than 98% of the tin in the ore is in cassiterite, 1–2% of the tin is in the sulphide minerals of tin.

Streaks, inclusions and clusters of grains of cassiterite, arsenopyrite and lellingite represent ore mineralization. The size of individual grains of ore minerals is up to 5.0 mm, the apparent thickness of clusters is up to 60 mm. Cassiterite is observed in the form of xenomorphic grains, their clusters; crystals of a prismatic, pyramidal-prismatic appearance are less often observed (Fig. 1). The grains size of cassiterite varies from 0.01 to 2.0 mm, with a predominance in the range of 0.1–0.5 mm, the thickness of clusters of 3–5 mm, sometimes can reach up to 10.0 mm.

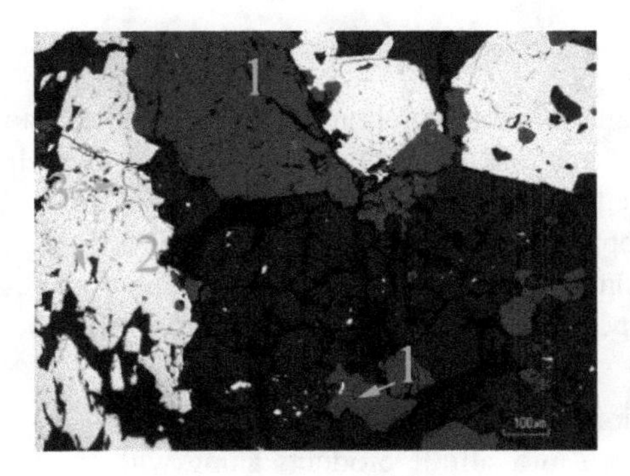

Fig. 1. Micrograph of cassiterite (1) in the intergrowth with arsenopyrite (2), lellingite (3).

According to the mineralogical analysis, crushed ore mainly consists of fragments of greisen, quartz, topaz, tourmaline, feldspar and mica. Ore minerals (arsenopyrite, lellingite, bornite and chalcopyrite) in crushed ore are in the form of fragments of crystals and grains of irregular shape. Sometimes they are found in intergrowths with each other and with rock-forming minerals.

Cassiterite is observed in the form of irregular shape grains, as well as in intergrowths with rock-forming minerals, less often with arsenopyrite or lellingite. The results of the study of the disclosure of cassiterite grains are shown in Table 1.

Table 1. Disclosure of cassiterite grains in crushed ore

Size class, mm	*Cassiterite grains and cassiterite – rich intergrowth, %*	*Cassiterite-containing intergrowth, %*	*Total*
−2+1	96,9	3,1	100,0
−1+0,5	98,9	1,1	100,0
−0,5+0,315	98,2	1,8	100,0
−0,315 +0,25	99,8	0,2	100,0
−0,25+0,125	99,8	0,2	100,0
−0,125+0,071	99,9	0,1	100,0
−0,071 +0,040	100,0	0,0	100,0
−0,040+0,0	100,0	0,0	100,0

According to Table 1, in ore crushed to a particle size of less than 2 mm, cassiterite grains are mainly in the form of free grains. The maximum number of intergrowths (3.1%) is noted in the size class −2 + 1 mm. The complete disengagement of cassiterite grains from intergrowths is achieved in size less than 0.071 mm.

Tin sulphide minerals (stannin, mawsonite, stannoidite) are associated with later copper mineralization. These minerals are in close intergrowth with chalcopyrite and bornite. Their size usually does not exceed 0.1 mm.

4 Conclusions

Tin minerals, which are valuable components of the studied samples, have a significantly higher degree of contrast of gravitational properties, which can be used for their primary concentration Together with cassiterite and stannine, wolframite, arsenopyrite with lellingite and copper sulphides will be extracted into the primary gravity concentrate. To obtain tin concentrate that meets the requirements for raw materials, arsenopyrite and copper sulfides can be extracted from the rough concentrate using flotation methods.

During the flotation process, the sulfide minerals of tin (stannin, mawsonite, stannoidite) are extracted into sulfide products along with the minerals of copper and arsenic. To extract tin sulfides minerals, selective flotation with preliminary fine grinding of sulfide products will be required.

Differences in the magnetic susceptibility of minerals, with the use of magnetic separation, wolframite, biotite and tourmaline can be distinguished.

Studying the contrast of the physical properties of ore minerals provided an idea of the mineral composition of the products of primary gravity enrichment, which will make it possible to develop an optimal scheme for finishing operations.

Acknowledgements. The Irkutsk State University, individual research grant № 091-18-231, supported this work.

References

Bashlykova TV (2005) Tekhnologicheskie aspekty racional'nogo nedropol'zovaniya: Rol' tekhnologicheskoj ocenki v razvitii i upravlenii mineral'no-syr'evoj bazoj strany [Technological aspects of rational subsoil use: The role of technological assessment in the development and management of the country's mineral resource base]. Moscow^ MiSIS (in Russian)

Danilov UG, Grigoryev VP (2017) Problemy i perspektivy razvitiya olovyannoj promyshlennosti Rossii. Gornaya promyshlenńost 5(135):83–87 (in Russian)

3

Nanotechnologies in Mineral-Geochemical Methods for Assessing the Forms of Finding of Gold, Related Elements, Technological Properties of Industrial Ores and their Tails

R. Koneev[1(✉)], R. Khalmatov[2], O. Tursunkulov[2], A. Krivosheeva[1], N. Iskandarov[2], and A. Sigida[1]

[1] National University of Uzbekistan, Tashkent, Uzbekistan
ri.koneev@gmail.com
[2] Centre for Advanced Technology, Tashkent, Uzbekistan

Abstract. Seven mineralogical-geochemical types of ores have been identified in the gold orogenic deposits of Uzbekistan: /Au-W/Au-Bi-Te/Au-As/Au-Ag-Te/Au-Ag-Se/Au-Sb/Au-Hg/. Non-industrial are Au-W and Au-Hg. For each industrial type, certain gold compounds are characteristic: maldonite, Au-arsenopyrite and Au-pyrite, petzite, physhesserite, petrovskaite, aurostibite, which form regular micro- nanoensembles with the corresponding minerals Bi, Te, Se, S, As, Sb. They are direct indicators of the type and technological properties of ores. Pyrite and arsenopyrite are preserved in the processing wastes of Au-Bi-Te and Au-As ores, in which gold (901‰), maldonite (Au_2Bi), headleyite (Bi_7Te_3) and others were detected, 300–700 nm in size. Waste is suitable for the secondary extraction of gold.

Keywords: Nanomineralogy · Gold · Ores · Tails · Properties · Technologies

1 Introduction

The Republic of Uzbekistan is one of the leading gold mining countries in the world. Such large deposits as Muruntau, Myutenbay, Kokpatas, Zarmitan and others (Kyzylkum-Nurata ore region) are known. As a result of many years of production, the mining and processing industry faces the problem of transition from oxidized ores with large "free" gold to the processing of refractory ores with "invisible" gold. One of the effective approaches to the assessment of such ores is the use of nanotechnology, nanomineralogy and nanogeochemistry (Hochella 2002; Koneev et al. 2010).

2 Methods and Approaches

The section of mineralogy that studies the conditions of formation, the physicochemical properties of natural compounds, the size of which, at least in one dimension, enters the nanoscale (10^{-6}–10^{-9} m) is called nanomineralogy. There are nanominerals that are formed by the "down-top" technology, from atoms to chemical compounds, and nanoparticles that are formed by the "top-down" technology during dispersion. The anomalous properties of nanominerals and nanoparticles are determined by a significant increase in the specific surface energy. As a result of "size effects", nano-objects acquire high chemical, catalytic, sorption activity and other properties. "Noble", chemically inert gold in the nanostate becomes active and is found in ores not only in the form of native, but also in the form of a compound with various elements up to sulfur and oxygen. When studying ores, the main focus is on ore minerals. Non-metallic minerals, carbonaceous material rather negatively affect the enrichment processes, making it difficult to crush, float or adsorb gold. The study of mineralogy and nanomineralogy is important due to the fact that minerals of different composition and size differ in their ability to crush, float, or cyanate. The studies used an electron probe microanalyzer Superprobe JXA-8800R, Carl Zeise Oxford instrument (SEM-EDX); ICP MS 7500 Agilent Technologies.

3 Results and Discussion

The gold ore deposits of the Kuzylkum-Nurata region of Uzbekistan are confined to the South Tien Shan orogenic belt. Placed in the "black" shale, terrigenous, carbonate, volcanogenic and intrusive rocks. Tails of ore enrichment stored in long-term facilities were also studied. It has been established that in the orogenic deposits of the Kyzylkum-Nurata region, seven mineral-geochemical ore types are developed: /Au-W/Au-Bi-Te/Au-As/Au-Ag-Te/Au-Ag-Se/Au-Sb/Au-Hg/. Each type differs in the ratio of non-metallic and ore (sulphide) minerals, but primarily in the composition of micro-nanoensembles of minerals, compounds and gold fineness. In the studied deposits, the leading industrial ones are Au-Bi-Te, Au-As, Au-Ag-Se, Au-Sb (Table 1). In addition to native gold, electrum and kustelite, maldonite, Au-arsenopyrite, Au-pyrite, petzite,

Table 1. Types, composition of ores of gold and gold-silver deposits of Kyzylkum-Nurata region

Deposits	Main ores types	Leading mineral of ore	Micro-nanomineral ensemble	Gold compounds
Gold-quartz type **Muruntau Myutenbay Triada**	Au-Bi-Te, bismuth-telluride; Au-As, pyrite-arsenopyrite;	Quartz, albite, arsenopyrite, pyrite, scheelite	Pilsenite, hedleyite, ingodite, tsumoite, joseite, tetradymite, kobellite, bismuth	Native, maldonite (Au_2Bi), Au-arsenopyrite Au-pyrite
Gold-sulfide-quartz type **Zarmitan Urtalik Gujumsay**	Au-Bi-Te, bismuth-telluride; Au-As, pyrite-arsenopyrite; Au-Sb, antimony-sulfoantimony;	Quartz, carbonate, albite, arsenopyrite, pyrite, scheelite	Pilsenite, хедлейит, tsumoite, joseite, kobellite, bismuth, boulangerite, jamesonite, tetradymite	Native, electrum, maldonite (Au_2Bi), aurostibite ($AuSb_2$), Au- arsenopyrite Au- pyrite
Gold-sulfide type **Amantaytau Daugyztau Kokpatas**	Au-As, pyrite-arsenopyrite; Au-Sb, antimony-sulfoantimony;	Quartz, carbonate, pyrite, arsenopyrite, antimonite	Jamesonite, bournonite, boulangerite, zinkenite, tetradymite	Native, electrum, aurostibite ($AuSb_2$), Au- pyrite
Gold-silver type **Kosmanachi Visokovol'tnoe Adzhibugut**	Au-As, pyrite-arsenopyrite; Au-Ag-Se, sulfosal-selenide;	Quartz, carbonate, pyrite, arsenopyrite, galena	Silver, freibergite, pyrargyrite, polybasite, naumannite, aguilarite	Electrum, kustelite, petrovskaite (AuAgS), fishesserite (Ag_3AuSe_2)

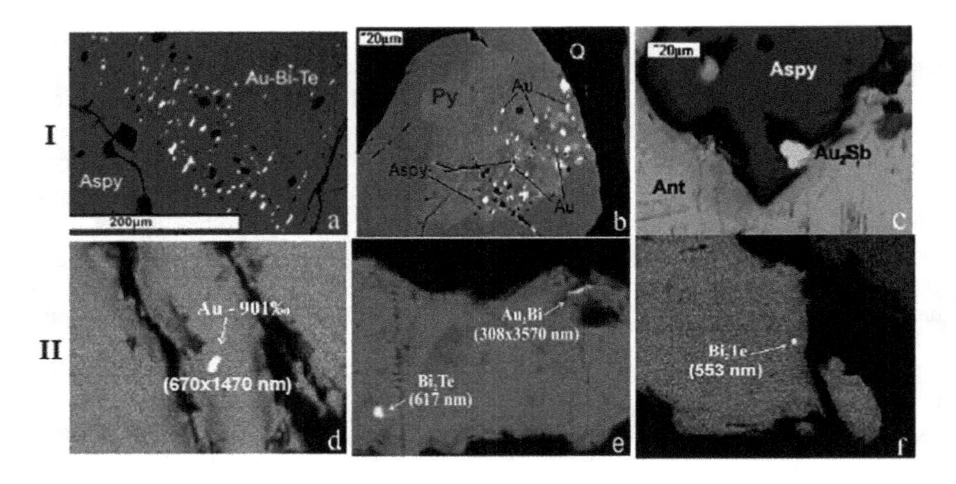

Fig. 1. Micro-nanomineral forms of gold release and its compounds in various types of ores (I) and tails (II). a - maldonite, Bi-tellurides in arsenopyrite; b - gold, arsenopyrite in arsenic pyrite; c - aurostibite with arsenopyrite in antimonite; d - nanogold in pyrite; e - maldonite and hedleyite and arsenopyrite; f - hedleyite in arsenopyrite.

calavertite, fishesserite, petrovskaite, aurostibite were detected in ores. Gold release forms and its micro- nanoanesmble in ores in Fig. 1, row I. The study of Au-Bi-Te and Au-As ore tails showed that some of the sulfide minerals still remain, in which the nanoforms of gold, its compounds and associated bismuth tellurides are preserved 300–700 nm in size (Fig. 1, row II).nts.

4 Conclusions

As a result of the research the following were established: - Seven mineral-geochemical types of ores were identified in the orogenic deposits of the Kyzylkum-Nurata region of Uzbekistan: /Au-W/Au-Bi-Te/Au-As/Au-Ag-Te/Au-Ag-Se/Au-Sb/Au-Hg/. Industrial types are /Au-Bi-Te/Au-As/Au-Ag-Se/Au-Sb/ ores possessing different technological properties; - Gold is represented by its native type, electrum, cyustelite, maldonite, petzite, aurostibite, fishesserite, petrovskaite, Au-arsenopyrite, Au-pyrite, forming micro- nanoensembles with the Bi, Te, Sb, Se, As, S minerals; - Pyrite and arsenopyrite remain in the enrichment tails of Au-Bi-Te and Au-As ores, in which nanoinclusions of gold, maldonite and bismuth tellurides are found. Tails are recyclable; - Nanotechnological methods allow to increase the efficiency and unambiguity of exploration, assess the prospects for hidden and unconventional deposits, as well as assess the technological properties of ores and enrichment tails for their complete processing.

Acknowledgements. The authors are grateful to the Navoi Mining and Metallurgical Combinate for their assistance in research.

References

Hochella M (2002) There's plenty of room at the bottom: nano-scince in geochemistry. Geochem Cosmochem Acta 66(5):735–743

Koneev RI, Khalmatov RA, Mun YS (2010) Nanomineralogy and nanogeochemistry of ores from gold deposits of Uzbekistan. Geol Ore Deposits 52(8):755–766

4

Microtomographic Study of Gabbro-Diabase Structural Transformations under Compressive Loads

L. Vaisberg[1(✉)] and E. Kameneva[2]

[1] «Mekhanobr-Tekhnika» REC, St. Petersburg, Russia
gornyi@mtspb.com
[2] Petrozavodsk State University, Petrozavodsk, Russia

Abstract. Transformations of the porous space structure for gabbro-diabase under compressive loads is studied. Quantitative assessment of respective structural parameters for the pore space is ensured through the application of X-ray computer microtomography, enabling visualization of the internal three-dimensional structure of each sample and a detailed quantitative analysis of the pore space structure for both separate sections and the entire sample volume. Differences in the number, sizes, shapes, connectivity and spatial distribution of pores are established. It is shown that, when the sample is destroyed, the structure of the pore space in its fragments is transformed as follows: intracrystalline pores are partially closing, with the simultaneous emergence of new pores of large capillary sizes, concentrated in the cracks. In terms of their structure, these cracks represent a system of interconnected pores containing micron-size mineral particles.

Keywords: Computer microtomography · Gabbro-diabase · Pore space

1 Introduction

In the current theory for the mechanism of disintegration of rocks, destruction is a process that develops in time. The formation and development of microscopic failures begin upon application of a loading force, either dynamic or static, and resume for the entire period the rock remains under load until fracture (Zhurkov 1980; Krivtsov 2007).

The existing works on rock destruction generally identify the most probable structural elements, along which the destruction processes tend to develop. These are pores, tiny fractures, intergrowth boundaries of mineral phases and intergrain boundaries. In this regard, porosity, understood as the sum of all cavities enclosed in the rock, including pores, pore channels, and tiny fractures, becomes a useful feature, linking the strength of a rock with the defects in its structure. The physical or total porosity of a rock, determined by calculation using known values of mineral and bulk densities, enables only indirectly assessing the transformations occurring in the rock microstructure during destruction. Unbiased data may only be obtained if the structure of the pore space is taken into account, including its dimensions, pore shapes, connectivity and spatial orientation (Romm 1985).

The purpose of this work is to study the transformations occurring in the microstructure of a rock during destruction. X-ray computer microtomography was selected as the method for the quantitative assessment of respective pore space structural parameters of rocks at the microscopic level.

2 Methods and Approaches

X-ray tomography (X-ray micro-CT) is a non-destructive method for studying the internal structure of solid materials, based on the dependence of the linear coefficient of X-ray radiation attenuation on the chemical composition and density of the substance analyzed. Computer processing of shadow projections obtained by x-ray scanning of samples allows visualizing the internal three-dimensional structure of each sample and performing a detailed analysis of its morphometric and density characteristics both at separate sections and throughout the entire sample volume, obtaining quantitative values of respective parameters. The non-destructive nature of the method is an important advantage of X-ray microtomography, as it enables subsequent application of the same samples for other types of analysis, in particular, for establishing their strength characteristics.

A gabbro-diabase sample of cylindrical shape (L = d) without visible defects was prepared for the studies. The gabbro-diabase sample was characterized by a massive texture and a uniform medium-grained structure. The size of mineral grains was up to 2–3 mm.

The experiment included X-ray tomography of the gabbro-diabase sample with identification of the following parameters of the pore space: dimensions, shape, volume, specific surface, pore connectivity, spatial orientation, distribution, and pore density, both in separate tomographic Sections (2D system) and in the entire volume of the sample (3D system). The sample was then subjected to axial loading using a manual hydraulic press until fracture. The loading force was 192 MPa, the lateral pressure was 1 atm, at t = 20 °C, the loading time was 121 s. The resulting fragment was then subjected to x-ray tomography.

The tomographic studies were carried out using SkyScan-1172 (Belgium) with resolutions of 0.5 to 27 μm. In the experiments, the samples were carefully oriented on the table with respect to the optical axis of the instrument; the tube was supplied with 100 mA and 100 kV; the X-ray power was 90 W. The pixel size at the maximum magnification (nominal resolution) was 3.9 μm, which generally allows identifying pores of 4 μm or larger. The table with the sample was rotated by 360° in 0.25 increments. The subsequent reconstruction work was performed using Nrecon, CTan, and CTvol SkyScan software.

3 Results and Discussion

The original sample contains crystals of 0.2–2.0 mm, represented by plagioclase (41.5%), actinolite (47.0%), quartz (3.6%), sphene (5.7%) and biotite (2.2%).

The porosity of the initial sample is 0.7%. The pores in the sample volume are unevenly distributed. Plagioclase crystals have the highest porosity that is 4.1 to 4.9 times higher than the value for the entire sample (2.9–3.4% vs 0.7%). In quartz, actinolite and biotite crystals, only individual pores are observed.

The concentration of pores in the original sample is 84.55 mm^{-3}.

The largest pore size is 32–34 μm. In quantitative terms, pores of up to 10 μm prevail. The largest pore size is 32–34 μm with their median value of 5.6 μm. In quantitative terms, pores of up to 4–5 μm (80.7%) prevail in this group.

In the fragment after fracture, the total pore concentration increases to 163.9 mm^{-3}, super-capillary pores of 179 to 180 μm are observed, and the porosity increases to 1.8% (Table 1).

Table 1. Results of microtomography (3D-system)

Parameters	Values	
	Original sample	Fragment after sample fracture
Pore fraction in sample volume, %	0.7	1.8
Concentration of pores, mm^{-3}	84.55	163.9
Largest pore size, μm	32–34	178–180
Pore connectivity, %	2.81	28.88

Pore sphericity decreases from small pores to large pores. Pores of 4–5 μm are close to spherical in shape. Pores of 110–180 μm have elongated shapes, with their sphericity not exceeding 0.15 to 0.3.

It can be seen on the tomographic sections (Fig. 1) that the fragments formed upon sample fracture contain incomplete cracks (not leading to the formation of fracture surfaces) with the length of L = 7–8 mm and width of h = 40–150 μm (for main cracks) and of h = 1–10 μm (for feather cracks).

The study of the structure of the cracks shows that these are linear-plane sections consisting of interconnected cavities containing micron-size mineral particles (Fig. 2).

The availability of these cracks satisfactorily explains the increase in pore connectivity. In the original sample, the pore connectivity is 2.81%; in the fractured fragment, it is an order of magnitude higher (27.88%). The increase in connectivity indicates the association of small pores into larger pores.

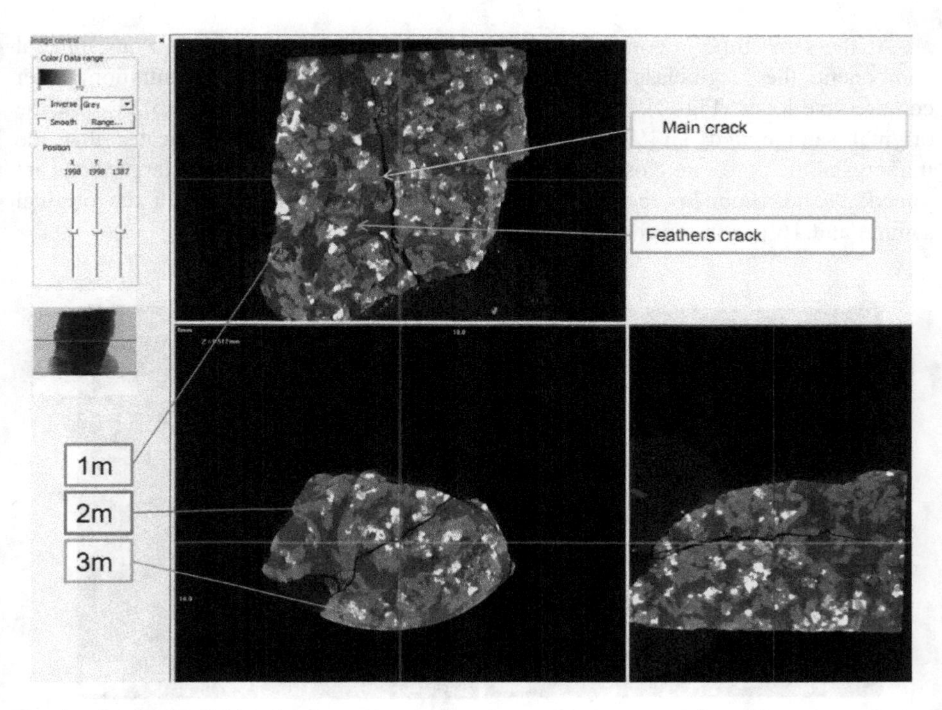

Fig. 1. Cracks in Gabbro-Diabase sample fragments (2D System) 1 m—*Plagioclase*; 2 m—*Actinolite*, 3—*Sphene*

Fig. 2. Structure of cracks

At the same time, a comparative tomographic study of the most porous mineral component, the plagioclase crystal, shows a decrease in pore concentration under compressive loads (Fig. 3). The plagioclase crystal porosity values are 3.23% for the original sample and 1.11% for the fractured fragment, which indicates that the intracrystalline pores are closing. Not only the finest pores, but also the larger ones are joined: the maximum pore size in the plagioclase crystal is 32 μm in the original sample and 16 μm in the fragment.

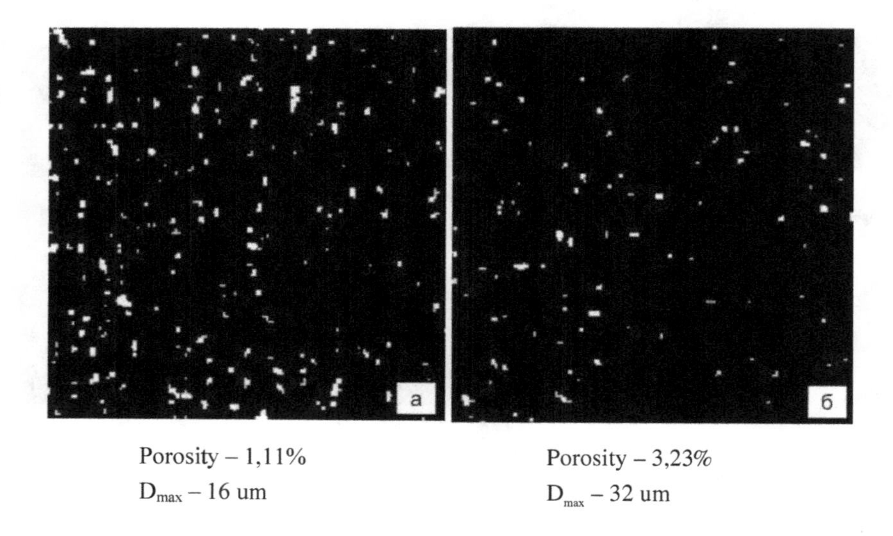

Fig. 3. Pores (white) in Plagioclase crystal in original sample (a) and in its fractured fragment (b) in 2D system

The increase in the number of pores in the volume of the fragment formed after fracture, with the closing of intracrystalline pores, suggests that the newly formed pores are mainly concentrated in the cracks. In terms of their structures, these cracks represent a system of interconnected pores containing micron-size mineral particles.

4 Conclusions

When a gabbro-diabase sample is fractured under the action of a compressive load, the structure of the pore space is transformed in the fragments formed as follows: intracrystalline pores are partially closing, with the simultaneous emergence of new pores of large capillary sizes, concentrated in the cracks.

Acknowledgements. The study was supported by the grant of the Russian Science Foundation (project No. 17-79-30056).

References

Krivtsov AM (2007) Deformation and fracture of solids with microstructure, Fizmatlit, Moscow

Romm YeS (1985) Structural models of pore space of rocks, Nedra, Leningrad

Zhurkov SN (1980) On the question of the physical nature of the strength. J Solid State Phys 22 (11):13–15

5

Th/U Relations as an Indicator of the Genesis of Metamorphic Zircons

Y. Pystina(✉) and A. Pystin

Institute of Geology Komi SC UB RAS, Syktyvkar, Russia
pystina@geo.komisc.ru

Abstract. Having studied polymetamorphic complexes of the Urals, including its northern part, for many years we have collected material on the basis of which we attempted to make generalizations concerning both the morphology of zircons and their geochemical features, allowing the mineral to be used in the reconstruction of specific metamorphic events and the interpretation of geochronological data (Pystina et al. 2017; Pystina and Pystin 2002). In recent years we also obtained new results on the morphology and geochemistry of zircons from granitoids in the northern part of the Subpolar Urals (Pystina and Pystin 2002). Together, this made it possible to compare different morphological types of magmatic and metamorphic zircons.

Keywords: Polymetamorphic complexes · Geochronological · Morphology · Geochemistry of zircons

1 Introduction

Precambrian formations, especially Pre-Riphean, which underwent metamorphism, as a rule, experienced it repeatedly, i.e. were polymetamorphic. Accordingly, the zircons of the newly formed, or transformed from the previously existing ones, in the course of these events should have acquired some new properties, expressed in changes in the morphology of the crystals, the internal structure, the geochemical composition. That is what we see in zircons from various polymetamorphic complexes of the Urals, including those located in its northern part: Nyartin, in the Subpolar Urals, and Harbey in the Polar Urals, where up to five morphological types of this mineral are distinguished (Pystina and Pystin 2002).

2 Results and Discussion

Detritic zircons (type 1) determine the metamorphic affiliation to one or another source formation. Zircons of the "soccer ball" type or, as is customary in the Urals, after Krasnobayev (1986), call them "granulitic" (type 2), and also "migmatite" (type 4) fix several age levels of occurrences of high-temperature rock transformations. Zircon of irregular shape like "cauliflower" (type 3) is typical for rocks metamorphosed under conditions that do not exceed the low to medium stages of the amphibolite facies. In more high-temperature conditions, it is found in the rocks of the basic series. The reason for the emergence of such intricate forms of zircon may be the absence or deficiency of silicate melt. Therefore, when P-T reaches the conditions of metamorphism sufficient for the development of migmatization processes, such zircon can continue to crystallize only in metamorphic mafic composition, for which, as is known, the migmatization temperature is higher. Opaque zircons (type 5) are associated with the manifestation of medium temperature diaphtoresis. The internal structure of all morphotypes of zircons is characterized by the presence of nuclei of irregular or rounded shape, in the "migmatitic" type, oscillatory zonality is usually noted, the "granulitic" type is the most homogeneous. Among the isolated morphotypes of zircons in polymetamorphic complexes, the "granulite" and "migmatite" types prevail. If, by morphological features and internal structure, the zircons of polymetamorphic complexes are surely divided into morphotypes, which can be associated with certain metamorphic events or processes, then the geochemical composition of the scattered elements does not make a clear separation. On the one hand, this is due to the extremely low content of the elements themselves, most often zircon is enriched only in Hf, Y, U, P. C, on the other hand, the nature of the distribution of these elements in the crystal, their quantitative variations do not give grounds to separate some zircon morphotypes from others. Although in some cases it is possible. E.g., in "migmatite" zircons (type 4) from the gneiss of the Harbay Complex, the distribution of Hf decreases from the center of the crystal to the edge, while in other morphotypes of zircons from the same rocks it increases. At the same time, in the gneisses of the Nyartin Complex in all the selected morphotypes, including the "migmatite" type, the content of Hf increases from the center of the crystal to the edge. The same picture, according to our data, is observed in zircons from the rocks of the metamorphic complexes of the Southern Urals: the Alexandrov and Ilmenogorsky (Selyankin block). The content and distribution of the scattered elements according to the data available today with morphological types of zircons are not clearly correlated. In this regard, the Th/U ratio deserves interest, which for magmatic zircons are, as a rule, > 0.5 (Skublov et al. 2009), and for metamorphic, significantly lower – 0.1–0.3 (Pystin and Pystina 2015a), according to Rubatto < 0.07 (Rubatto 2002), although according to other data it can be > 0.5, for example, 0.73 in zircons from the eclogite of the Maksyutov Complex (Pystin and Pystina 2015b). But, despite some rebounds in Th/U values, the average indices for metamorphic zircons, as well as for magmatic ones, are fairly consistent. In the metamorphic zircons of the gneisses of the Nyartin Complex, obtained from 9 crystals, they vary from 0.02 to 0.39, but among these values, two are essentially out of the general picture – 0.75 and 0.68. Such values of the Th/U ratio correspond to zircons of magmatic origin. In our case, these are zircons of prismatic habitus, which we have isolated into the "migmatite" type. The isotope age, obtained by the U-Pb SHRIMP-II method, is 503 ± 8 Ma and 498 ± 8 Ma, respectively (Pystin and Pystina 2008). Their formation was associated with the metamorphism of the amphibolite facies and the associated granitization. Therefore, the conditions, under which this morphotype was formed, were similar to the conditions of granite formation, hence we assume high Th/U ratios.

An even clearer picture is obtained for zircons of the "granulitic" type from the Alexandrov Complex in the Southern Urals, where the Th/U ratio varies from 0.23 to

0.31, which is quite consistent with the metamorphic zircons. This is confirmed by the isotopic age of all the crystals, which is close and approximately 2.1 Ga (Pystin and Pystina 2015b, Pystina and Pystin 2002).

Our studies show that accessory zircons from rocks of different granitoid complexes in the northern part of the Subpolar Urals, which have different geological positions and isotopic age, differ in the set of morphotypes, their quantitative ratios and geochemical features (Pystina et al. 2017). Recently we studied morphological features of zircons from the granites of Nikolayshor (PR1), Kozhim (PF2–3), Badyayu (RF3–V) and Yarota (RF3–V) Massifs. Accessory zircons are very diverse in form, character of zonality, presence of inclusions, color, degree of metamictism and other features. All the main morphological types of zircons according to I.V. Nosyrev (Nosyrev et al. 1989) were observed: zirconium, hyacinth, spear-shaped, torpedo-shaped and citrolite. All of the above morphological types can relate to the generation of zircons of either synpetrogenic or superimposed genetic types.

In the granites of the Nikolashor Massif hyacinth, spear-shaped, torpedo-shaped, and detrital zircons were found. Granitoids of the Kozhimsky, Badyayusky and Yarotsky Massifs are characterized by the presence of three morphotypes of zircons, but if in the granitoids of the Badyayu and Yarota Massifs they are similar (zircon, hyacinth and torpedo-shaped morphotypes), while in the rocks of the Kozhim Massif, they are pattern, torpedo-shaped, torrent, and the beginning of torpedo-shaped morphotypes. Common to the granitoids of all the massifs is one morphotype – torpedo. The spear-shaped zircon is found only in the rocks of the Nikolashor Massif. There are also detrital zircons, which are absent in the granitoids of other massifs. Granites of the Kozhim Massif are distinguished from other granitoids by the presence of zircon of the citrolite morphotype. The presence of this type of zircons is a sign of the metasomatic (or metamorphic) processing of rocks. In contrast to the granites of the Badyayu and Yarota Massifs, they lack hyacinth type zircons.

The Th/U ratio in zircons from the granitoids of the northern part of the Subpolar Urals – Nikolashor, Kozhim, Khatalambo-Lapchin and Lapchavozh Massifs, is on average 0.73; 0.61; 0.51; 0.79, respectively. These values are maintained and observed in all zircons of the studied granitoids. In some cases, in zircons from the granitoids of the Khatalambo-Lapchin Massif, the Th/U values are knocked out of the overall picture, amounting to 0.22 and 0.15, which is not at all characteristic of magmatic zircons. If we take into account that the age of these zircon crystals obtained by the U-Pb SHRIMP-II method is 703.9 $\pm$ 8 Ma and 795 $\pm$ 41 Ma, and the rest of the zircons is 550–580 Ma, we can assume that the formation of ancient zircons The early stages of granite formation and the increased Th/U ratio are explained by the subsequent metamorphism of early generation granites.

3 Conclusions

Thus, we have to state the validity of the fact that "the only obvious systematic difference between the magmatic and metamorphic zircon is the Th/U ratio ..." [Hoskin and Schaltegger 2003, p. 48]. It allows not only to distinguish between magmatic zircons from metamorphic, but taking into account the morphological features of individual crystals and isotopic age dating, it is more reliable to restore the history of the formation of specific metamorphic and igneous complexes.

Acknowledgements. This work was supported by the Basic Research Program of the Russian Academy of Sciences No. 18-5-5-19.

References

Hoskin PWO, Schaltegger U (2003) The composition of zircon and igneous and metamorphic petrogenesis. Rev Miner Geochem 53:27–62

Krasnobayev A (1986) Zircon as an indicator of geological processes. Nauk, Moscow, 152 p

Nosyrev IV, Robul VM, Esipchuk KE, Orsa VI (1989) Generation analysis of accessory zircon. Nauka Press, Moscow 203 p

Pystin A, Pystina J (2015a) The early precambrian history of rock metamorphism in the Urals segment of crust. Int Geol Rev 57(11–12):1650–1659

Pystin AM, Pystina YI (2008) Metamorphism and granite formation in the Proterozoic-Earlypaleozoic history of the formation of the Polar Ural segment of the earth's crust. Lithosphere (6):25–38

Pystin AM, Pystina YI (2015b) The Archean-Paleoproterozoic history of metamorphism of rocks of the Ural segment of the earth's crust. Works of the Karelian Research Center of the Russian Academy of Sciences, no 7, Geology of Dokembriya, pp 3–18

Pystina YI, Denisova YV, Pystin AM (2017) Typomorphic signs of zircons as a criterion for the dissection and correlation of granitoids (by the example of the northern part of the Subpolar Urals). Bulletin of the Institute of Geology of Komi Scientific Center of the Ural Branch of the Russian Academy of Sciences, no 12, pp 3–15

Pystina Y, Pystin A (2002) Zircon chronicle of the Ural Precambrian. UrD RAS Press, Ekaterinburg, 167 p

Rubatto D (2002) Zircon trace element geochemistry: partitioning with garnet and the link between U-Pb ages and metamorphism. Chem Geol 184:123–138

Skublov SG, Lobach-Zhuchenko SB, Guseva NS, et al (2009) Distribution of rare-earth and rare elements in zircons from miaskite lamproites of the Panozersky complex of central Karelia. Geochemistry (9):958–971

6

New Approaches in X-ray Phase Analysis of Gypsum Raw Material of Diverse Genesis

V. Klimenko(✉), V. Pavlenko, and T. Klimenko

Belgorod State Technological University named after V G Shukhov, Belgorod, Russia

klimenko3497@yandex.ru

Abstract. Modern software was used to conduct X-ray phase analysis of gypsum raw material of different genesis: gypsum from several deposits, citrogypsum, phosphogypsum, vitamin gypsum, chemically pure gypsum, hydration products from different gypsum and anhydrite cementing and composite materials on their bases. Two criteria for ranging gypsum raw material with the account of genesis and structural characteristics and predicting properties of gypsum bindings and materials based on them were suggested: structure sophistication criteria (K_g) and identity criteria (K_i). They were calculated by the results of X-ray phase analysis of calcium sulphate.

Keywords: Gypsum · X-ray phase analysis · Structure sophistication criteria · Identity criteria · Gypsum texture and structure

1 Introduction

There are convincing theoretical and experimental data about physical-chemical processes of gypsum dehydration, their structural defects dominating role, an origin and transformation mechanism of these defects during calcinations, at the same time there are no data about influence of the gypsum raw material formation on these processes. Thermodynamic characteristics of gypsum with different structure are rather close (Kelly et al. 1948; Reference… 2004) and it is difficult to study gypsum raw material deformations according to them. There are data (Gordashevski 1963) about application of differential thermal and x-ray analysis methods for these purposes. To characterize gypsum from different deposits P.F. Gordashevskiy (Gordashevski and Sakhno 1963) suggested identity criteria – the value received by division of reflexes intensivities difference at $2\Theta = 20^\circ$ and 23° by their sum. On the other hand, gypsum crystal lattice defects can be characterized by crystallinity degree, which is determined by reflex with d = 2.81 Å in the X-ray patterns (Boldyrev 1983). Crystallinity degree diagnostics is possible in relation to diffraction reflection doublet intensivity and by crystallinity index determination. Some authors suggest studying crystal lattice defects by studying X-ray diffraction line broadening (Ginje 1961). Literature analysis shows wide researches in this sphere as well as absence of practical standards. In connection with this the purpose of this work was to find new criteria allowing ranging gypsum raw material with the account of genesis and structural characteristics at designing composite materials.

2 Materials and Methods

As raw materials we studied gypsum of Shedok, Baskunchak, Novomoskovsk, and Peshelansk Deposits; man-induced gypsum – chemically pure gypsum, citrogypsum, vitamin gypsum, Voskresensk phosphogypsum, synthetical gypsum; gypsum thermal treatment and rehydration products, composite materials based on gypsum binders, multiphase gypsum binders (MGB). To receive composite materials, we used crashed glass withdrawals (CGW) and iron-ore concentrate of Lebedinsk mining and processing enterprise (OMC). X-ray phase analysis of the studied species was done with X-ray diffraction meter DRON-4 by powder pattern method. Grain-size analysis of powdery material was done by laser granulometry method with a MicroSizer 201 installation.

3 Results and Discussions

The X-ray phase analysis of the initial raw material showed that intensity and area of the main reflexes depend on its genesis. Intensity (L) and reflex area (S) at interplanar spacing (d) 7,628 Å are strongly dependent on gypsum raw material. Gypsum raw material genesis influences somewhat lower the intensity and reflex area at d = 4.291 and 3.069 Å. Reflex intensity at d = 3.809 Å almost independent of the gypsum raw material nature. The received data evidence that the studied gypsum samples have equal crystal lattice dimensions.

To determine the gypsum raw material genesis influence on its structure and crystal lattice defects several studies of the initial gypsum raw material crystallinity degree and its products of the heat processing have been studied. Here indexes were suggested calculated by reflex area difference division by their sum. The calculations were done for reflexes with the most intensity such as: 7.628 Å; 4.291 Å; 3.809 Å; 3.069 Å; 2.880 Å; 2.687 Å.

We found that most indexes slightly depended on raw material genesis. Certain dependencies were observed in changes of two indexes chosen for further work. The first index was structure sophistication index (K_g), and the second was identity index (K_i). K_g, was determined by X-ray phase analysis results, as ratio of reflex area difference division at 2Θ = 29.09° (d = 3.069 Å) and 2Θ = 31.08° (d = 2.876 Å) by their sum, a K_i as a ration of reflex area difference division at 2Θ = 20.68° (d = 4.291 Å) and 2Θ = 31.08° (d = 2.876 Å) by their sum.

$$Kg = \frac{S_{29.09} - S_{31.08}}{S_{29.09} + S_{31.08}} \qquad K_i = \frac{S_{20.68} - S_{31.08}}{S_{20.68} + S_{31.08}}$$

For natural gypsum K_g = 0.42–0.46, and for production induced gypsum was 0.560–0.903 (Table 1). The greater the value K_g, the more sophisticated crystal structure was and gypsum raw material was less stable.

We think that K_i value depends on gypsum raw material micro-assembly dimensions. Powdery-material grain-size composition determined by laser granulometry analysis proves a finely crystalline structure of phosphogypsum and citrogypsum.

Table 1. X-ray phase analysis of gypsum of different genesis

№	Gypsum genesis	Index value	
		Ki	Kg
1	Chemically pure gypsum	0.612	0.660
2	Shedok deposit gypsum	0.362	0.440
3	Baskunchak deposit gypsum	0.448	0.420
4	Novomoskovsk deposit gypsum	0.448	0.450
5	Peshelansk deposit gypsum	0.215	0.450
6	Vitamin gypsum	0.746	0.903
7	Voscresensk phosphogypsum	0.404	0.560
8	Citrogypsum 1	0.312	0.640
9	$CaCl_2$ and $(NH_4)_2SO_4$ gypsum	-	0.749
10	Anhydrite cement (1,5% $(NH_4)_2SO_4$ + 0,5%$CuSO_4$)	0.277	0.169
11	Composite material based on MGB and CGW remains (Klimenko et al. 2013)	0.286	0.156
12	Composite material: 70% $CuSO_4$ II + 30% LC 70% Г-4 + 30% LC (Klimenko et al. 2018)	0.370 0.270	- -

Crystal size of production induced chemically pure gypsum and vitamin gypsum is higher. They can be referred to large grained gypsum. By value K_i Shedok natural gypsum can be referred to finely crystalline raw material (finely tessellated oriented structure). Baskunchak natural gypsum has medium tessellated chaotic structure, and according to value K_i it takes intermediate position.

Hence, for large grained gypsum K_i = 0.612–0.746, for finely crystalline gypsum K_i = 0.312–0.488. Granulometry of produced gypsum changes unevenly. There are fractions, which number depends greater on genesis, and there are fractions, which number depends less on raw material genesis. To analyze the influence of gypsum heat treatment parameters and amount of residue hydrate water on value K_g and K_i we used Baskunchak natural gypsum. The results show that K_g of calcium sulphate hydration products can be both greater and less K_g of natural gypsum. Gypsum produced during heat treatment products hydration with hydrate water 11.00–14.42 mass%, has sophistication structure index value greater than that of natural gypsum. The smallest value Kg, irrespective of heat treatment parameters is of calcium sulphate hydration products with hydrate water amount 3–4 mass%. Calcium sulphate hydration products having value K_g close to K_g of the initial gypsum, have maximum strength. The amount of residue hydrate water of these products is equal 1.0–1.5 mas. %. By value K_i (0.39–0.45) the produced gypsum can be classified as fine- or medium crystalline structure.

Analysis of identity index value suggests that calcium sulphate hydration products with the amount of hydration water 10–15 mass% have bigger crystal micro cluster size and hydration products β-$CaSO_4{\cdot}0.5H_2O$. β-centrifuged hemihydrates of calcium sulphate and soluble anhydrous plaster (β-$CaSO_4{\cdot}III$) have smaller size of gypsum crystal micro clusters (K_i = 0.35–0.43).

4 Conclusion

A method and parameters (structure sophistication index (K_g) have been suggested and identity index (K_i)) allowing ranging gypsum raw material with the account of genesis and structural characteristics and forecast properties of gypsum cementing components and materials on their basis. It has been determined that for natural gypsum K_g = 0.42–0.46 and for produced gypsum K_g = 0.560–0.903. Value K_i depends on micro clusters sizes of gypsum raw material structure. For large grained gypsum K_i = 0.612–0.746. for finely crystalline gypsums – 0.312–0.488. At producing gypsum cementing agents it is necessary to observe that their K_g is closer to value K_g of natural gypsum and K_i is in the range 0.38–0.41.

Acknowledgements. The work is realized within the framework of the Program of flagship university development on the base of the Belgorod State Technological University named after V.G. Shukhov. using equipment of High Technology Center at BSTU named after V.G. Shukhov.

References

Boldyrev VV (1983) Experimental methods in mechanochemistry of solid inorganic matters. Science, Novosibirsk, 36 p

Ginje A (1961) Crystal X-ray filming. State Publishing House of Physic-Mathematical Literature, Moscow, 188 p

Gordashevskiy PF, Sakhno ZA (1963) About some properties of gypsum raw material of different crystalline structure, no 26. Collection of works/SRILCM, Moscow, pp 25–27

Gordashevskiy PF (1963) Thermal and X-ray phase gypsum analyses results. Constr Mater (12):28–30

Kelly N, Suttard D, Anderson K (1948) Thermodynamic properties of gypsum. BTI MCMPI, Moscow, pp 38–42

Klimenko VG, Kashin GA, Prikaznova TA (2018) Plaster-based magnetite composite materials in construction. In: IOP Conference Series: Materials Science and Engineering, vol 327, p 032029. https://doi.org/10.1088/1757-899x/327/3/032029

Klimenko VG, Pavlenko VI, Gasanov SK (2013) The role of pH medium in forming binding substauces on base of calcium sulphate. Middle-East J Sci Res 17(8):1169–1175

Reference book. Gypsum materials and articles (production and application) (2004) Under general editorship of professor, doctor of sciences A.V. Ferronskaya. Construction Universities Association Publishing House, Moscow, 485 p

7

Properties and Processing of Ores Containing Layered Silicates

A. Gerasimov(✉), V. Arsentyev, and V. Lazareva

REC Mekhanobr-tekhnica, St-Petersburg, Russia
gerasimov_am@npk-mt.spb.ru

Abstract. The solid mineral raw materials of the sediment deposits containing the argillaceous variations such as the coal and the potash ores are the most multi-tonnage solid minerals mined and processed in the territory of the Russian Federation. Both types of the deposits have relatively simple geology aspects with large and medium-sized solids and as a rule, the most often, with the regular bedding which are characterized by the fitchery thickness and the internal structure. Such deposits are very difficult for mining and processing. The thermal modification of this of ores allows changing the physico-chemical properties of its components in such a way that their further processing - grinding, separation, transportation and storage can be carried out without the use of process water or significantly reduces its consumption.

Keywords: Argillaceous minerals · Hydrochemical modification · Thermochemical modification · Water saving

1 Introduction

Many deposits of minerals are associated with the presence of layered silicates such as the clay, mudstones, argillaceous slate - highly hydrophilic mineral varieties that have the ability to swell in the water. This ability to self-disperse creates the great difficulties both for the extraction and for the processing of such ores. Since the processing of most minerals is associated with the use of wet processes, where in contact with water the swelling layered silicates are dispersed to a particle size of less than 50 μm, but due to their high hydrophylic property they form bonds that promote the formation of so-called structured suspensions having increased viscosity. In this case, the structured suspensions make it difficult to perform the separation operations as the classification, the gravity and magnetic separation, the flotation and thickening and filtration operations (Arsentyev et al. 2014).

In order to eliminate the influence of swelling silicates on the treatment processes it is necessary to use the diluted suspensions with the solids content of 10–20% while the usual solids content in processing suspensions is 35–50%. This causes the increase in 3–4 times the consumption of water during the processing of ores containing the swelling sheet silicates (Zhdanovich et al. 2011).

The most large-tonnage processed commercial mineral containing the swelling silicates are the coal and the potassium salts. Considering these minerals from the point of view of their advanced processing, it is important to note the similar processing characteristics:

- comparatively high content of the useful component which is 60–80% for the bituminous coal and 20–30% for potassium ores;
- polydispersity of the impregnations of the useful component is from 0.5 to 10 mm;
- relatively low density of the constituent rocks is from 1.4 to 2.3 g/cm^3;
- the presence of argillaceous material such as the layered silicates swelling in the water.

2 Methods and Approaches

Coal, sylvinite and kaolin were selected for the dry processing study.

The presented coals are dark gray, dense, laminated and fractured with the matte surface in the place of spalling. The samples of the coal are brittle and under a slight stress are destroyed along the microfracture (Gerasimov et al. 2016).

The sylvinite sample is a rock salt-sylvinite-carnalite formation with the varying ratio of the main rock-forming minerals represented mainly by readily soluble congeries and individuals of the chloride class (Arsentyev et al. 2017a, b).

The kaolin ore is represented by the kaolin argillites mainly gray, dark gray, greenish-gray, less often yellowish-ochreous and sorrel, occasionally milky-gray colors cured during the compaction, dehydration and cementation processes. The structure is massive often the breccia structure and splintery, sometimes the oolitic structure, here and there the indistinctly-banded. The structure is the pelitic, aleuropelitic, aleuro-psammito-pelitic, occasionally oolitic (Arsentyev et al. 2017a, b).

3 Results and Discussion

The thermal treatment of the high-ash coals (low-temperature pyrolysis reaction) modefies the physicochemical properties of the minerals that make up the ash reducing their ability to swell in water and increasing the magnetic susceptibility which makes it possible to significantly intensify the processes of their magnetic-flotation processing to produce the high-quality solid fuels. In the coal, under the influence of temperatures in the process of the coal pyrolysis, the initial fracture structure is transformed into the cracked-pore structure with the increase in the number and dimensions of the poriness, vesicles and cracks. The sample of the coal subjected to medium-temperature pyrolysis reaction compared to the initial coal has 20% lower apparent density and the specific energy consumption of crushing is 20–30% lower.

The specific ash content per unit of the heat of combustion indicates that the combined dry processing of coal allows reducing the specific ash content by 1.3–1.6 times i.e. the formation of the ash and slag during the combustion is reduced.

It has been established that the thermal treatment of the sylvinite ore; containing the argllous-carbonate insoluble minerals in the range of 100–500 °C practically does not affect the structure of salting minerals of the halite and the silvinite but it significantly modifies the structure of minerals entering into the insoluble fraction which has the positive effect on the flotation of these ores and the process of sedimentation of the slurry fractions in the flotation tails.

For reducing the energy costs for the heat treatment of sylvinite ores it has been proposed to use microwave treatment.

The hydrothermal treatment of kaolin raw material in the autoclave can significantly reduce the viscosity of kaolin suspensions and further efficient processing to produce the concentrates for the production of alumina by using the wet screening process for the suspended matters with the relatively high density.

4 Conclusions

An analysis of international practice has shown that fresh water is becoming the most scarce resource for the mining and processing industry, which leads to a significant increase in the cost of water supply and circulation systems, with their growth rates being four to five times higher than the growth rates of the ore mining output.

Studies of the problems of reducing water consumption in mineral processing should be combined with energy consumption assessments due to the close relationship between water and energy consumption values within a single processing system of any plant. Most often, a technology that ensures water saving requires higher energy consumption and vice versa.

The presence of layered silicates with mobile crystal lattices in the mined ore significantly complicates both its wet and dry processing.

A promising solution, aimed at efficient use of resources in processing of ores containing layered silicates, implies the inclusion of thermal and hydrothermal modification of such ores as the primary stages of the process chain. These modifications would ensure a reduction or elimination of the negative impact of such materials currently experienced in ore preparation, separation and dewatering and in the storage of processing products.

The solution requires additional energy inputs, but ensures significant savings in other resources.

Acknowledgements. Financial support was provided by the Russian Scientific Fund (project 18-17-00169).

References

Arsentyev VA, Vaisberg LA, Ustinov ID (2014) Directions of creation of low-water technologies and devices for the processing of finely divided mineral raw materials. Obogashenie Rud (5):3–9

Arsentyev VA, Gerasimov AM, Kotova EL (2017a) Thermochemical modification of sylvinite ore using the microwave heating. Obogashenie Rud (6):3–8

Arsentyev VA, Vaisberg LA, Ustinov ID, Gerasimov AM (2016) Prospects for reducing water use in coal-processing. Mining J (5):97–100

Arsentyev VA, Gerasimov AM, Mezenin AO (2017b) Study of technology of kaolin processing using hydrothermal modification. Obogashenie Rud (2):3–9

Gerasimov AM, Dmitriev SV (2016) Combined technology of dry coal processing. Obogashenie Rud (6):9–13

Lyskova MY (2016) Geoecology in the modern construction of enterprises for the extraction and enrichment of potassium-magnesium salts, News of Tula State University. Geosciences (4):39–49

Mozheiko FF, Potkin TN (2008) Physico-chemical basis of processing of high-arcilla off-balance sylvinite ores. Vesti, National Academy of Sciences of Belarus, Series of Chemical Sciences, no 4, pp 25–32

Titkov SN et al (2013) The technology of dry crushing of potash ore to flotation size. In: IX Congress of enrichers of the CIS countries. Collection of materials, vol 2. Digest MISA. M. 2013, pp 578–583

Zhdanovich IB et al (2011) Influence of heat treatment of saliferous arcilla on the structural-geological properties of their dispersions. Vesti, National Academy of Sciences of Belarus, Series of Chemical Sciences, no 3, pp 113–117

8

Ore Mineralogical Study of Cerattepe Au-Cu (±Zn) VMS Deposit

İ. Akpınar[1(✉)] and E. Çiftçi[2]

[1] Department of Geological Engineering, Faculty of Engineering and Natural Science, Gumushane University, Gumushane, Turkey
hiakpinar@gumushane.edu.tr

[2] Department of Geology, Faculty of Mines, Istanbul Technical University, Istanbul, Turkey

Abstract. The Cerattepe mine, one of the volcanogenic massive sulfide deposits in northeastern Turkey, is hosted within the late Cretaceous volcanic, intrusive and sedimentary rocks. Deposit's main ore body contains high-grade massive copper sulfides and a gold-silver and barium rich oxide zone, characterized by dense alteration stages, is situated on top of it. Replacement, cataclastic, breccia, dissemination, dendritic, concentric growth, colloform, and framboidal textures were identified. Pyrite, sphalerite, marcasite, chalcopyrite, bornite, galena, tennantite-tetrahedrite, gold, covellite, digenite, chalcocite, cuprite and cubanite constitute the mineral paragenesis where quartz, calcite and barite account for the gangue minerals. Limonite, hematite, lepidocrocite, malachite, azurite and jarosite developed in the oxidation zone.

Keywords: Eastern Pontide · Cerattepe · VMS deposit · Framboidal pyrite · Bird's eye-texture

1 Introduction

The eastern Pontide orogenic belt of Turkey is an important segment of the Tethyan-Eurasian Metallogenic Belt. This belt carries a special importance in metallogeny of Turkey and hosts numerous VMS deposit. (Akıncı 1984; Çiftçi and Hagni 2005; Güven 1993; Revan et al. 2013; Zaykov et al. 2006; Yiğit 2005). The most of the eastern Pontide VMS deposits show some similarities in many aspects to the Kuroko deposits of Japan (Çiftçi and Hagni 2005; Ciftci 2000; Pejatoviç 1979). The Cerattepe Deposit is a Kuroko-type VMS deposit located in late Cretaceous age volcanic, intrusive and sedimentary rocks. It is distinguished by an unusual basal zone of high-grade copper sulfides and an overlying Au-rich oxide zone from the other VMS deposits of northeastern Pontides.

2 Methods and Approaches

A total of 46 samples representing oxide (14), sulfide (28) and stringer zones (4) of Cerattepe ore deposit were collected from drill cores, underground audits and surface outcrops. Polished sections were prepared and Nikon Eclipse LV100 reflected light microscopy was employed for examination. The ore minerals and the paragenesis were identified on the basis of their petrographical features and their textural relationships, respectively. Electron Probe Micro Analysis (EPMA) and Secondary Electron Microscopy-Energy Dispersive Spectroscopy (SEM-EDS) were used for chemistry of sulfide minerals.

3 Results and Discussion

The mineral paragenesis (Fig. 1) of Cerattepe VMS deposit comprises of pyrite, sphalerite, marcasite, chalcopyrite, bornite-idaite, galena, covellite, chalcocite, cubanite, cuprite including sulfosalts (mainly tennantite and lesser tetrahedrite) gold, silver, arsenopyrite and bournonite.

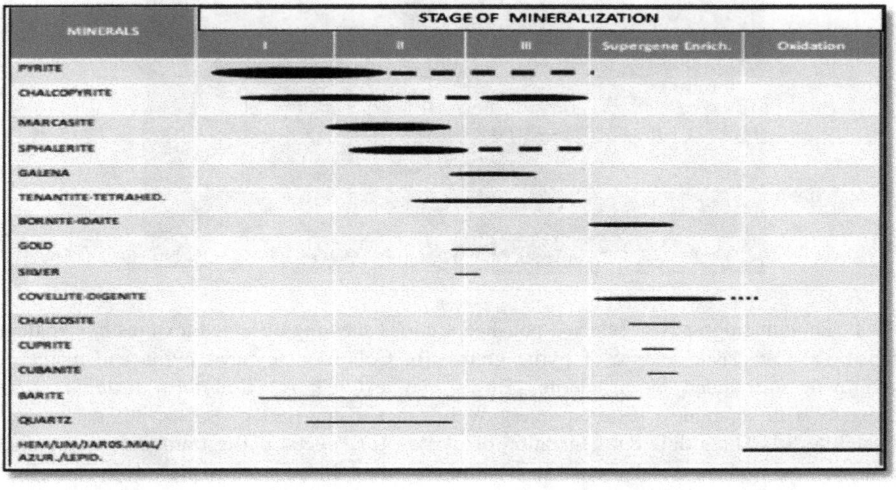

Fig. 1. Ore mineral paragenetic sequence of Cerattepe Au-Cu (±Zn) VMS deposit

Gangue minerals include barite, quartz, gypsum, anhydrite and calcite. Hematite, limonite, lepidocrocite, malachite and azurite, and jarosite are the oxidation minerals. Four generation of pyrite is specified. Pyrite I is represented by colloform-concentric textured (Fig. 2) grains formed from initial solution reached the seafloor from the chimney. The colloform textured pyrites are progressively overgrown later coarse crystalline grains, which are the second generation pyrites.

These are later extensively altered to marcasite. The framboidal-pelletal textured collomorphic grains, seen sometimes coeval with sphalerite and galena, are the third generation pyrites. The fourth generation pyrites are small sized euhedral, subhedral grains observed on transition zone of sulfide ore- footwall rock and veinlets of stringer zone.

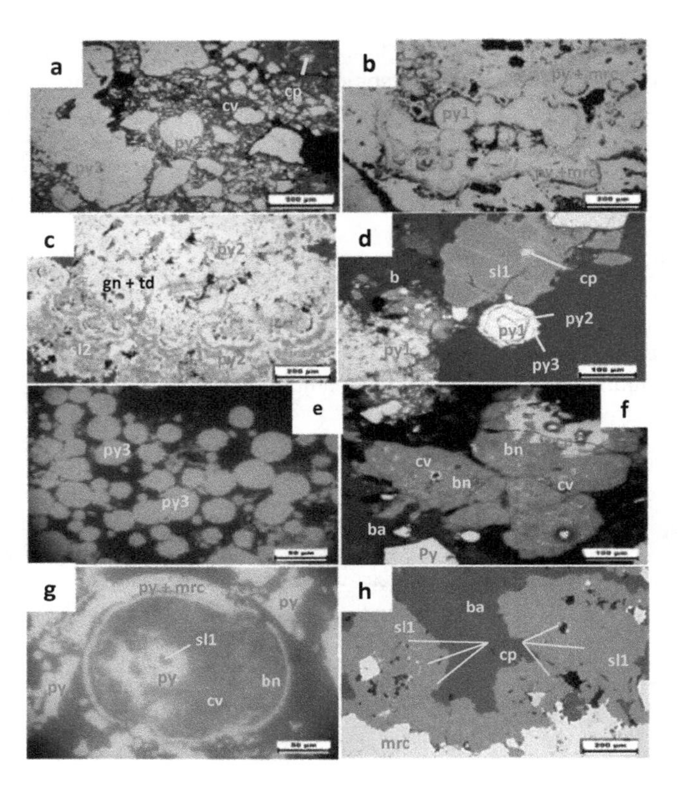

Fig. 2. Photomicrographs of selected polished sections showing observed textures in Cerattepe deposit. (a) Cataclastic texture of pyrite grains. (b) Colloform textured pyrite and marcasite intergrowth surrounded by sphalerite "Bird's eye texture"; (c) Colloform textured pyrite aggregate with a spongy textured intergrowth zones comprised of sphalerite, galena and tetrahedrite; (d) Three different generation of pyrites, (e) Pelletal pyrite framboids with minor sphalerite and galena in the cracks; (f) Bornite and chalcopyrite replaced by covellite; (g) collomorphically banded sulfides in melnikovite pyrite; (h) Dissociation texture of chalcopyrite in sphalerite replaced by marcasite

Most of the minerals are small - fine grained, and the larger grains of the major minerals are in the order of 100–800 μm up to 1.2–2 mm in size and some rare pyrite grains have a size of 5 mm. Most of the minor and trace minerals are much smaller, typically in the order of 1–20 μm in size. Majority of ore minerals are anhedral with the exception of pyrite, quartz, barite and some sphalerite and galena occur as euhedral to subhedral crystals. Observed ore textures are dissemination and veinlet textures in the stockwork and siliceous ore zones, whereas replacement, overgrowth, concentric and colloform textures become prevalent in the massive ore, particularly in the center of main deposit.

Banded textures of black and yellow ore are seen with polymetallic sulfides. In the outer part of the main ore body, at lateral zones, clastic or fragmental ore textures are present.

4 Conclusions

It is concluded that the ore mineral assemblage and textures observed in Cerattepe VMS deposit are comparable to those of other VMS deposits occur in north eastern Pontide region, Kuroko deposits of Japan and also comparable modern sea-floor–seamount VMS deposits in the world.

Acknowledgements. This study was financially supported by BAP Project unit (No: 39615) of Istanbul Technical University in Turkey. We are thankful to Mr. Ünsal Arkadaş and staff of Etibakır A.Ş. in Artvin, Turkey for their courteous support for the field work.

References

Akıncı ÖT (1984) The geology and the metallogeny of the Eastern Pontides (Turkey). In: International world geological congress abstracts, pp 197–198

Ciftci E (2000) Mineralogy, paragenetic sequence, geochemistry, and genesis of the gold and silver bearing upper cretaceous mineral deposits, Northeastern Turkey. Ph.D. Thesis, University of Missouri-Rolla, Rolla, MO, USA (unpublished)

Çiftçi E, Hagni RD (2005) Mineralogy of the Lahanos deposit a Kuroko-type VMS deposit from the eastern Pontides (Giresun, NE Turkey). Geol Bull Turk 48(1):55–64

Güven İH (1993) Artvin Kafkasör Sahası Maden Jeolojisi Raporu No 2600, MTA, Ankara. metallogeny. Ore Geol Rev 147–179

Pejatovic S (1979) Metallogeny of the Pontid type massive sulphide deposits, Ankara, 98 p

Revan MK, Genç Y, Maslennikov V, Ünlü T, Delibaş O, Hamzaçebi S (2013) Original findings on the ore-bearing facies of volcanogenic massive sulphide deposits in the eastern black sea region (NE Turkey). Bull MTA 147:73–89

Yigit O (2005) Gold in Turkey - a missing link and metallogenic features of the in Tethyan

Zaykov V, Novoselov K, Kotlyarov V (2006) Native gold and tellurides in the Murgul and Çayeli volcanogenic Cu deposits (Turkey)

9

Applied Mineralogy for Complex and Profound Mineral Processing

V. Chanturiya(✉) and T. Matveeva

Institute of Comprehensive Exploitation of Mineral Resources,
Russian Academy of Sciences, Moscow, Russia
vchan@mail.ru

Abstract. The main methodological parameters and the modes of investigation of samples containing micro- and nano-sized reagent phases are determined. Using scanning laser microscopy and an original method for analyzing the surface relief, a quantitative assessment of the adsorption layer of the collector reagent on Au-sulfide minerals was performed, the proportion of the molecular form of adsorption and the retention agent fixing strength was calculated. The theoretical basis for choosing the reagent mode for selective flotation of multicomponent ores was developed. The action mechanism has been revealed and the prospect of using novel selective collectors and environmentally friendly plant reagents for extracting non-ferrous and noble metals from complex sulfide ores has been substantiated.

Keywords: Reagent · Adsorption · Nanoparticles · Mineral · Microscopy · Flotation

1 Introduction

At present, modern mining and ore dressing plants have been facing a number of serious challenges in processing low-grade complex ores and technogenic resources, increasing demands of high quality metal concentrates and ecologically safe methods of beneficiation. In those conditions, the tasks of making mineral processing more complete and comprehensive and of creating highly effective technologies come to the fore. These should be based on the intensification of the existing methods and on creating new methods of mineral extraction from hard-to-enrich ores and from technogenic deposits by using the newest achievements of the fundamental sciences. The transition to the new strategy of primary processing is only possible on technological-mineralogical evaluation of raw minerals.

A modern complex of high-resolution physical methods allows to investigate the composition, structure and properties of geomaterials at the micro and nano-level, including:

- Identify micro and nano-sized particles of noble metals and surface natural and artificial nano-formations on minerals;
- Experimentally substantiate the structural, phase and chemical transformations of minerals under various energy methods of influence;

- To substantiate the choice and mechanism of interaction of reagents with noble metals during flotation of complex ores of complex material composition;
- Investigate the structural, phase and chemical transformations of sulfides and rocks in heap leaching processes.

2 Methods and Approaches

Research methods are optical, confocal laser, analytical electronic, scanning probe microscopy, UV-spectrophotometry of reagent solutions, flotation of minerals. The KEYENCE scanning laser microscope with the surface analysis module VK-9700 enables making a non-contact measurement of the roughness of the surface of minerals and thus determining the height and size of the new formations obtained as a result of interaction with the reagents. The electronic microscope with energy-dispersive microanalyzer LEO-1420 VP INCA-350 allows determining the elemental composition of micro- and nanophases of reagents on the surface of minerals. The analysis of the surface of minerals before and after a contact with reagent solutions was carried out on polished sections made in the form of polished plates $10 \times 10 \times 2$ mm in size.

3 Results and Discussion

In IPKON RAS within the framework of the scientific school of academician V. A. Chanturiya a complex of theoretical and experimental studies on the research and testing of new classes of complex-forming reagents – collectors and modifiers for flotation extraction of non-ferrous and noble metals from refractory gold-bearing oreshave been made. To analyze the processes of physicochemical effects of flotation reagents on micro and nano inclusions of gold and platinum on the surface of sulfides, the authors first developed and improved methods for obtaining samples of mineral complexes that mimic natural sulfides containing "invisible" or submicron gold or platinum. The new highly effective reagents developed in IPKON RAS are modified diethyldithiocarbamate DEDTC and diisobutyldithiophosphinate DIFM, dithiazine derivatives MTX and dialkylpyrylmethane DAM. Those reagents showed the effect for extracting micro- and nanoparticles of noble metals while enriching the mineral raw materials of complex composition owing to formation of poorly water soluble gold compounds and their selective adsorption on gold containing sulfides, improving the flotation selectivity compared to traditional xanthate collector (Chanturiya 2017; Chanturiya et al. 2016; Matveeva et al. 2017a; Matveeva et al. 2017b). Studies were carried out using a set of UV and IR spectroscopy methods (UV-1700 Shimadzu and Infralum FT-8), analytical scanning electron (ASEM) (LEO 1420VP with INCA Oxford micro 350) and laser (KEYENCE VK-9700) microscopy, X-ray phase analysis (X-ray diffractometry). The results made it possible to establish the conditions for the formation of an adsorption layer of new reagents on micro- and nanoparticles of noble metals and to ensure an increase in gold recovery during flotation (Fig. 1).

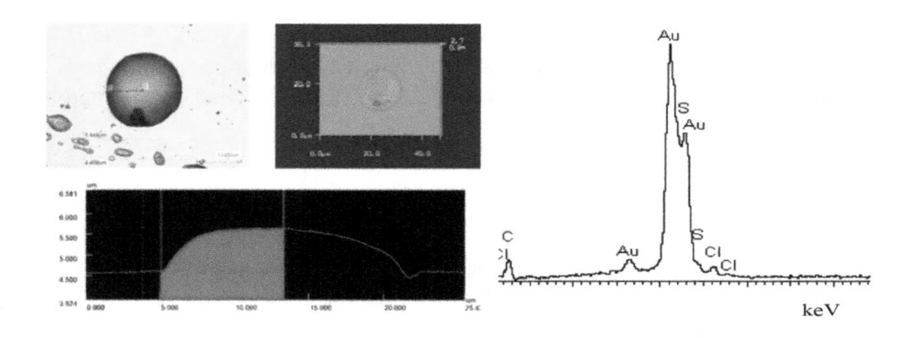

Fig. 1. Adsorbed layer of collector DEDTCm on the surface of Au-pyrite.

As the major losses, from 25% to 30% of valuable components (gold and minerals of the platinum group), are accounted for by the micro- and nano-size of the mineral particles whose concentration in the ore does not generally exceed 1,5–3,0 g/t, it was necessary to obtain samples imitating natural minerals. IPKON RAS has scientifically substantiated and offered procedures for artificially coating minerals with Au and Pt micro- and nano-particles, which made it possible to investigate the interaction mechanism for a new class of frothing agents with noble metals.

The introduction of the new agents both increased the extraction and improved the quality of the mineral concentrate in the mineral processing of rebellious ores of noble metals with a complex material composition, Table 1.

Table 1. Indicators of flotation of the low-grade sulfide ore of the Fyodorovo-Panskoye deposit when employing DEDTCm and ButX

Flotation products	βPd	βPt	εPd	εPt
ButX Conc.	3.88	1.02	73.97	72.33
Tails	0.21	0.06	26.03	27.67
Ore	0.7	0.19	100	100
DEDTCmConc.	10.14	4.24	82.23	85.31
Tails	10.18	0.06	17.77	14.69
Ore	0.93	0.38	100	100

The implementation of new technologies at the mining-enrichment works in Russia will make it possible to increase metal extraction by 10–15%, to obtain high-grade finished products that are competitive in the world market, to involve unpayable ores and technogenic raw materials into processing and considerably ameliorate negative effects on the environment in the mining industry regions.

4 Conclusions

The development of the mining sciences should be based on modern achievements of applied mineralogy and innovative development of comprehensive, economic and effective exploitation of mineral resources. The novel methods ensure both the higher level of mineral extraction, the higher grade of obtained valuable components and high level of ecological safety.

Acknowledgements. The authors are grateful to the Grant of the President of RF "Scientific School" of acad. V.A. Chanturiya and Program of Presidium of RAS P39.

References

Chanturiya VA (2017) Scientific substantiation and development innovative complex processing of mineral resources. Gornyi J 11:7–13

Chanturiya VA, Matveeva TN, Ivanova TA, Getman VV (2016) Mechanism of interaction of cloud point polymers with platinum and gold in flotation of finely disseminated precious metal ores. Miner Process Extr Metall Rev 37(3):187–195

Matveeva TN, Gromova NK, Minaev VA, Lantsiova LB (2017a) Modification of sulfide minerals surface and cassiterite by stable complexes Me–dibutylditiocarbamate. Obogashchenierud 5:15–20. https://doi.org/10.17580/or.2017.05.03

Matveeva TN, Ivanova TA, Getman VV, Gromova NK (2017b) New flotation reagents for extraction of micro- and nano-particles of noble metals from refractory ores. Gornyi J 11:89–93

10

Applied Mineralogy of Mining Industrial Wastes

O. Kotova[1(✉)] and E. Ozhogina[2]

[1] Institute of Geology Komi SC UB RAS, Syktyvkar, Russia
kotova@geo.komisc.ru
[2] FSBE "All-Russian Institute of Mineral Raw Materials" (VIMS), Moscow, Russia

Abstract. Recycling and disposal of mining wastes has been very important and considered as an independent task. Features of composition and structure of mining wastes, identified by a set of mineralogical analysis methods, allowed predicting their possible involvement in the secondary processing. This was illustrated by the example of metallurgical iron slags.

Keywords: Mining wastes · Technological mineralogy · Iron slags

1 Introduction

The intense development of the mineral resource complex inevitably leads to accumulation of significant amounts of waste, which negatively affect natural ecosystems. Therefore, the disposal and recycling of wastes is important at the national level and considered within framework of the priority direction of development of science and technology in the Russian Federation – "Rational environmental management". Today we can state the fact that recycling is an independent major task for our industry.

Mining and processing wastes are very diverse. They include overburden, host rocks, dry raw processing, off-balance sheet and non-standard ores, which composition and properties are not only close to their natural analogues, but usually used the same way. Processing wastes of metallurgical, chemical, heat power industries are more abundant. Slags, muds, ash slags and oil muds, burnt rocks, pyrite cinders, clinkers, dusts are significantly different from natural ores and rocks. They are characterized by variable granular composition, often a high dispersion, presence of amorphous formations, complex interrelationship of mineral and (or) technogenic phases, including presence of eutectic colonies or structures of decomposition of solid solutions, a small amount of one or more minerals, polymineral (polyphase) aggregates, presence of isomorphic minerals and polytypic modifications, secondary changes associated mainly with hypergenesis.

2 Methods and Approaches

The study of the composition, structure and technological properties of mining wastes is based on modern scientific, methodological, technical and instrumental support for the researches of technogenic raw and allows predicting its possible involvement in secondary processing, including elimination of environmental consequences of industrial processing (Ozhogina et al. 2018; Ozhogina et al. 2017; Chanturia et al. 2016; Yakushina et al. 2015; Ozhogna and Kotova 2019; Burtsev et al. 2018). Necessary and sufficient mineralogical information about the object is possible to get by a complex of mineralogical and analytical methods (optical and electron microscopy, X-ray, X-ray tomography, micro X-ray spectral analysis). For different types of wastes, an individual set of mineralogical analysis methods is used, which allows to obtain complete and reliable information, including information on the phase composition of technogenic formations, as well as the form of finding useful elements, granular composition, morphometric parameters, nature of localization of specific phases (Ozhogina et al. 2017).

Mineralogical study of mining wastes is carried out mainly according to the methodological documents developed for the analysis of natural mineral raw. Special methods of mineralogical analysis of industrial wastes do not exist. The study of such objects is of an interdisciplinary nature because of a reasonable combination of methods from various fields of science and adapted to solving mineralogical, technological and environmental problems.

3 Results and Discussion

The bulk of the processing wastes are slags. For example, iron-containing metallurgical slag is a loose material with dense lumpy aggregates and few octahedral crystals. More than 80% of the slag is represented by a thin material with a particle size of less than 0.2 mm. The main beneficial elements are iron (42.5%) and chromium (3.15%), which form their own mineral phases. Nickel (0.4%), associated with trevorite, and cobalt (0.08%), with unknown occurrence form, can possess industrial value.

A significant part of the slags (more than 75%) is formed by black magnetic material prone to artificial segregation, represented by spinelides forming a continuous isomorphic series spinel-magnetite-chromite. The main mineral is magnetite, which isolations are always non-uniform and contain numerous inclusions of non-metallic phases represented by olivine, pyroxene, mica, corundum, feldspar and glass (Fig. 1a). Sometimes the grains are surrounded by a fairly flat border of iron hydroxides and contain veinlets that do not extend beyond their boundaries (Fig. 1b). Two varieties of magnetite are noted; eutectic colonies are evidence of their simultaneous presence. This may be due not only to the closest intergrowth of the phases, or the oxidation process of a magnet and its partial transition to maghemite, but also continuous isomorphic substitutions of the ferrospinels, including heterogeneous −2 and 3 valence cations.

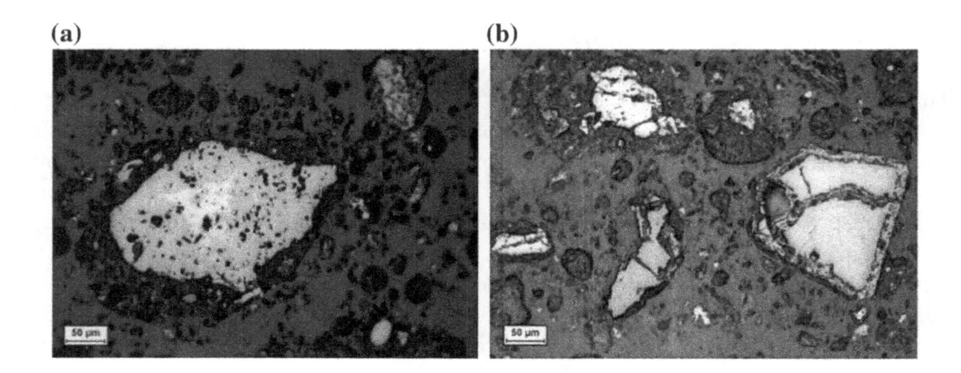

Fig. 1. A. Magnetite (light gray) with inclusions of non-metallic minerals, in the center of the grain is maghemite (white). B. Magnetite (light gray) with veins and border of iron hydroxides. Polished thin section, nicols are parallel

The features of the mineral composition and structure of iron-containing slags (phase composition, distribution of ore phases and aggregates, their morphostructural characteristics, heterogeneous structure of magnetite), identified by a complex of mineralogical analysis methods, suggest the expediency of chemical processing methods at recycling (Yakushina et al. 2015).

4 Conclusions

Mining wastes, being main types of technogenic raw, possess mineralogical characteristics (mineral and (or) phase composition, useful component occurrence form, morphostructural features and distribution pattern, real composition and structure) that determine the strategy and tactics of its secondary use:

(1) as initial raw without processing, for example, for extraction of valuable metals;

(2) as initial raw after additional processing to obtain material resources in the industry.

(3) as object of disposal.

References

Burtsev IN, Kotova OB, Kuzmin DV, Mashin DO, Perovsky IA, Ponaryadov AV, Shushkov DA (2018) The role of technological research in the development of the mineral resource complex of the Timan-North Ural region. Explor Prot Miner Resour 5:38–47 (in Russian)

Chanturia VA, Ozhogina EG, Shadrunova IV (2016) Tasks of ecological mineralogy during the development of the Earth's interior. Phys Tech Probl Dev Miner Resour (SB RAS) 5:193–196 (in Russian)

Ozhogina EG, Kotova OB, Yakushina OA (2018) Mining wastes: mineralogical features. Vestnik of IG Komi SC UB RAS, pp 43–49 (in Russian)

Ozhogina EG, Shadrunova IV, Chekushina TV (2017) The role of mineralogical studies in solving environmental problems of mining areas. Min J 11:105–110 (in Russian)

Ozhogina E, Kotova O (2019) New methods of mineral processing and technology for the progress of sustainability in complex ore treatment. In: 29th International Mineral Processing Congress IMPC 2018, Canadian Institute of Mining, Metallurgy and Petroleum, 2-s2.0-85059377649

Yakushina OA, Ozhogina EG, Khozyainov MS (2015) Microtomography of technogenic mineral raw materials. IG Komi Science Center, Ural Branch of the Russian Academy of Sciences. vol 11, pp 38–43 (in Russian)

11

Ag-Bearing Mineralization of Nevenrekan Deposit

E. Podolian[1,2(✉)], I. Shelukhina[1,2], and I. Kotova[2]

[1] "RMRL" Ltd. (Raw Materials Researching Laboratory), Saint Petersburg, Russia
podolyan@lims-lab.com
[2] Department of Mineral Deposits, Saint Petersburg State University, Saint Petersburg, Russia

Abstract. Nevenrekan deposit, located in Magadan region, Russia, is a perspective Au-Ag deposit with average silver content 445 ppm and gold 7.4 ppm. The study of the ores reveals the main Ag-bearing minerals: Au-Ag alloys, tetrahedrite-tennantite series and new phase – $Ag_8SnSe_2S_4$. New phase contains the main part of the silver of the deposit – 97.6%.

Keywords: Silver · Ag-sulphosalts · Ag-bearing minerals · Epithermal · Argyrodite

1 Introduction

The Nevenrekan Au-Ag deposit is located in the Severo-Even district of the Magadan region, Russia. The site is located within the central part of the Okhotsk-Chukotka volcanogenic belt. The main structural element is cretaceous-paleogenic Nevenrekan intrusive dome, which is crossed by quartz-adularia and quartz-carbonate veins with hydrothermal origin. These veins contain ores with Au-Ag mineralization. This work is motivated by mineralogical criteria for understanding ore-forming processes which are essential for future efficient exploration.

2 Methods and Approaches

Methods of research include: (1) chemical analyses and its interpretation (include X-Ray diffraction, atomic absorption analysis, inductively coupled plasma mass-spectrometry (ICP-MS) analysis for micro components and X-Ray fluorescence analysis for macrocomponents); (2) optical researches including petrographic research of thin sections and mineragraphic research of polish sections which are provided by electron microprobe analyses of chemical composition of ore minerals; (3) technological experiments (gravity и floatation methods of ores separation).

3 Results and Discussion

The main types of rocks at the deposit are clastolaval rhyodacites and ignimbrite rhyodacites, which are host rocks, and quartz-adularia veins, which are gangue rocks. Ore mineralization of Nevenrekan deposit consists of 4 ore associations which are characterized by different temperature of formation: 1. rutile-kassiterite; 2. pyrite-arsenopyrite; 3. sphalerite-chalcopyrite-galena-Au-Ag alloys-stannite series-$Ag_8SnSe_2S_4$-tennantite-tetrahedrite series with Ag; 4. hydrohematite. The most important paragenesis is the third one because it includes the main mineral concentrators of silver and gold. Technological and following chemical and mineralogical researches of beneficiaries demonstrate that the main concentrator of silver is sulphosalts, especially new phase $Ag_8SnSe_2S_4$, which contains 97.6% of all silver. According to crystallographic researches, phase $Ag_8SnSe_2S_4$ is orthorhombic and should belong to argyrodite group (Zhai et al. 2018). Nevenrekan deposit belongs to low-sulfidation type in classification of epithermal deposits (White and Hedenquist 1995).

4 Conclusions

The main Ag-bearing minerals at the Nevenrekan deposit are Au-Ag alloys, tennantite-tetrahedrite series and $Ag_8SnSe_2S_4$ sulphosalt of argyrodite group, which contain 97.6% of silver. The mineral composition of ores and the mineral balance of silver determine the prospects of flotation flow sheet for ore beneficiation.

Acknowledgements. We thank Polymetal Engineering Company for the ore samples and RMRL Ltd. for sponsorship of the participation in 14th International Congress for Applied Mineralogy.

References

Zhai D, Bindi L, Voudouris P, Liu J, Tombros S, Li K (2018) Discovery of Se-rich canfieldite, Ag8Sn(S,Se)6, from the Shuangjianzishan Ag-Pb-Zn deposit, NE China: a multimethodic chemical and structural study. Miner Mag 1–21

White NC, Hedenquist JW (1995) Epithermal gold deposits styles, characteristics and exploration. SEG Newsl 23:9–13

12

Gold Extraction to Ferrosilicium, Production of Foam Silicate from Processing Tails of the Olimpiada Mining and Processing Complex Gold Processing Plant

A. Sazonov[1(✉)], V. Pavlov[2], S. Silyanov[1], and E. Zvyagina[1]

[1] Siberian Federal University, Krasnoyarsk, Russia
Sazonov_am@mail.ru
[2] Special Design and Technology Bureau "Science", Krasnoyarsk, Russia

Abstract. The paper describes the studies of processing tails of the Olimpiada Mining and Processing Complex with the methods of chemical, mineralogical, electronic microscope, deep reductive melting with division of the melt into a silicate and metal parts. It is demonstrated that 85% of the processing tails consist of the oxides: SiO_2, CaO, Al_2O_3, Fe_2O_3, H_2O and CO_2. The distribution of gold and silver is provided by the size classes of the initial blend, after melting of which in reduction conditions re-distribution of gold to the metal phase of the melt occurs. The silicate part of the melt when released into water in the "thermal shock" mode forms a light porous X-ray amorphous material "foam silicate", which also serves as a resource for stable chemical composition for production of a wide range of import substituting ceramic materials.

Keywords: Pyrometallurgy · Processing tails · Ferrosilicium · Gold extraction

1 Introduction

The development of the methods of flotation, gravity and metallurgic processing of gold bearing ores and man-made products today is aimed at increase of the recovery of the valuable product (Pavlov 2005; Pavlov et al. 2015; Meimanova and Nogayeva 2014; Bogdanovich et al. 2013; Tselyuk and Tselyuk 2013; Algebraistova et al. 2017; Amdur et al. 2015). There are almost no technologies for comprehensive use of processing tails to produce an additional marketable product both for task of comprehensive use of processing tails for production of foamed resources for ceramic items production with associated gold recovery to the ferrosilicium matrix is vital.

2 Methods

The paper proposes the pyrometallurgic approach to the solution of the problem of comprehensive wasteless processing of man-made gold-bearing resources with the method of deep reductive melting with division of the melt into the deferrized silicate part with its further chilling in the thermal shock mode (Pavlov et al. 2015) and the metal part with associated re-distribution of gold in the ferrosilicium matrix. The specimens of the silicate part of the melt were prepared for study using the powder technology; the metal specimens were covered with epoxy resin with subsequent polishing. The chemical phase composition was studied with the use of an X-ray fluorescent S2 RANGER analyzer and the scanning electronic microscope (SEM) Hitachi S-3400N with EMF Bruker.

3 Samples

The input material for pyrometallurgic extraction of gold is represented by the tails of the gold processing plant of the Olimpiada Mining and Processing Complex, which is the leader in Russia in terms of the ore processing volumes and marketable gold production. The gold-antimony and gold-arsenic ores of the Olimpiada deposit have a complex mineral composition and are extremely refractory for gold recovery from them. Most of gold is in the form of thin dissemination in sulfides (Kirik et al. 2017; Novozhilov et al. 2014). No more than 50% of gold is extracted by direct cyanation. At the gold processing plant the flotation concentrates of the ores are exposed to biooxidation using the BIONORD® technology and subsequent leaching to release gold capsuled in sulfides.

The processing tails have loose consistence, mainly sand-aleuritic-clayey size, with the content of (−0.071) mm fraction of about 25%. The chemical composition is dominated by SiO_2, CaO, Al_2O_3, H_2O and CO_2, the share of which is about 90 wt.%. The concentrations of S are 0.69%, C—2.35%, Sb—0.11%, As—0.19%, and Ag — $<$ .1%. The gold grade in individual samples as per the fire assay data varies from 0.2 to 0.97 g/t, with the average metal grade in the bulk sample received for study of 0.6 g/t. The main minerals of tails are quartz, calcite and stratified silicates making 98–99%. Sulfides, oxides, hydroxides and sulfates of iron, arsenic, antimony and tungsten are the impurities. In addition to the native highcarat gold, aurostibite ($AuSb_2$) and jonassonite ($AuBi_5S_4$) are present in single grains. Copper, antimony and mercury impurities are noted in some gold particles. Native gold particles are less than 0.045 mm in size (90%). Most of gold is noted in the form of micron inclusions in sulfides, quartz, carbonate and micas. Successive chemical leaching in the tail material using the method (Antropova et al. 1980) identified about 20% of mobile gold forms (water-soluble, sorbed, ferri-forms) and sulfide and telluride forms $\sim$ 15%.

4 Results

In the process of the experiment the flotation tails sample with the weight of 400 g was mixed with lime ($Ca(OH)_2$) and brown coal. The blend was exposed to reductive melting at the temperature of 1500 1550 °C, in the process of which melt division into a metal and silicate parts occurred. The silicate part of the melt was poured into water with production of foamed amorphous material (foam silicate). The recovery of foam silicate was 150 g. The metal phase recovery was 26 g (4.4%). Phase composition of the metal aggregate (wt.%): ferrosilicum (FeSi) 82.8; xifengite (Fe_5S_3) 8.97; wustite ($Fe_{0.974}O$) 0.79; wollastonite ($CaSiO_3$) 4.24; calcic clinoferrosilite ($Fe_{1.5}Ca_{0.5}(SiO_3)_2$) and elemental iron (Fe) 0.49. Chemical composition of the produced foam silicate (wt. %) – SiO_2—43.7; TiO_2—0.7; Al_2O_3—7.79; Fe_2O_3—0.19; MgO—2.22; CaO—42.5; K_2O—1.5; SO_3—0.77; Cl—0.27, and metal phases: Si—23.5; Al—1.6; Fe—66.2; Mn —2.22; Mg 0.47; Ca—1.97; S—0.42; As—0.68; Cl—0.14; P—1.0; Co—0.35; V—0.28; Cu—0.21; Au—0.2; and Cr—0.18. The optical research of the metal alloy showed non-uniform aggregate composition. Six individualized metal phases of more complex composition have been identified in the iron and silicon alloy matrix. Gold is a part of the alloy consisting of (wt.%): Au—0.25–5.11; Sb—0.4–0.7; Sn—0.57–3.30; As—up to 9.47; Cu—4.51–32.07; Fe—59.8–33.8; Mn—7.2–1.33; Ga—0.24–9.38; and Si—14.53–12.96.

5 Conclusions

Therefore, silicate and metallic semi-finished products have been produced as the result of deep reductive melting of processing tails: (1) ferrosilicium, which is a gold collector; (2) foam silicate material as an additional product of the main production, can be used for production of ceramic materials for different purposes. The use of the method of pyrometallurgic processing of processing tails allows mitigating their adverse effect on the environment.

References

Algebraistova NK, Makshanin AV, Burdakova EA, Markova AS (2017) Processing of precious - metal raw materials in centrifugal devices. Non-ferrous Met 1:18–22

Amdur AM, Vatolin NA, Fyodorov SA, Matushkina AM (2015) Movement of disperse drops of gold in porous bodies and oxide melts during heating. Rep Acad Sci 465(3):307–309

Antropova LV, Shuraleva AZ, Farfel LF, Aizenberg FM, Priyemov GA (1980) Forms of gold occurrence in the rock. Explor Met Eng 136:5–21

Bogdanovich AV, Vasilyev AM, Shneyerson YaM, Pleshkov MA (2013) Gold extraction from stale processing tails of pyritic copper-zinc ores. Ore Process 5:38–44

Kirik SD, Sazonov AM, Silyanov SA, Bayukov OA (2017) Study of disordering in the structure of natural arsenopyrite with the X-ray diffraction analysis of polycrystals and nuclear gamma resonance. J Siberian Federal Univ Eng Technol 10(5):578–592

Meimanova ZhS, Nogayeva KA (2014) Study of flotation processability of stale tails from the Solton-Sary processing plant. Nauka i novye Tekhnologii 2:15–16

Novozhilov YI, Gavrilov AM, Yablokova SV, Arefyeva VI (2014) Unique commercial gold antimony deposit Olimpiada in upper proterozoic terrigenous deposits. Ores Met 3:51–64

Pavlov VF (2005) Physical bases for the technology of production of new materials with set properties on the basis of creation of the system for comprehensive use of man-made and barren resources. Siberian Branch of the Russian Academy of Sciences, Novosibirsk

Pavlov MV, Pavlov IV, Pavlov VF, Shabanova OV, Shabanov AV (2015) Features of processes of pyrometallurgic processing of polymetallic ores. Chem Benefit Sustain Dev 3:263–266

Tselyuk OI, Tselyuk DI (2013) Prospects of use of gold heap leaching for involvement into commercial development of stale tails of the Eastern Siberia gold processing plants. In: Proceedings of the Siberian Department of the Section of Earth Sciences of the Russian Academy of Natural Sciences, vol 1, no. 42 pp 103–110

13

Quality Assurance Support (QA/QC System) of Mineralogical Analysis

O. Yakushina(✉), E. Gorbatova, E. Ozhogina, and A. Rogozhin
FSBE "All-Russian Institute on Mineral Raw Materials" (VIMS), Moscow, Russia
vims@df.ru

Abstract. Mineralogical studies are an integral part of the exploration and development of solid mineral deposits, the effectiveness of which directly depends on the quality of the measurements. Moreover, five decades QA/QC system of mineralogical analysis (UKARM) for Russian geological survey is developed. Its specific features and tasks are discussed. The system provides to obtain complete, reliable, metrologically evaluated and legally valid information about the material composition and structure of rocks and ores. QA/QC of mineralogical analysis ensures the coordination of testing laboratories, starting with the resources of testing laboratories, stuffing at last. This system covers the entire process of mineralogical research, starting at the resources of testing laboratories as stuff and equipment, through the research procedure as the selection of testing object and its preparation for analysis, the accuracy of analysis rank, the testing method and technique, to metrology data and the quality assessment of the results obtained.

Keywords: Mineralogical analysis · Quality management · Assurance · Control system · Metrology · Reference materials

1 Introduction

Mineralogical analysis is an integral part of mineral deposits exploration and development; its effectiveness directly depends on the quality of testing (ISO/IEC 17025:2005; JORC Code 2012). The quality of analysis directly affects efficiency and reliability of the whole investigation. The main requirement for mineralogical research is to obtain reliable, complete, metrologically evaluated and legally valid information about the studied matter composition and structure, namely of rocks, ores and technologically processed products. The quality management system of mineralogical works ensures the coordination of testing laboratories, starting with the resources of testing laboratories, and ending with the processes occurring in them. Laboratory mineralogical studies with varying degrees of depth of mineral substance analysis are carried out at all stages of geological exploration from geological prospecting to operational exploration and development of mineral deposits. The main requirement for laboratory mineralogical studies is to provide all spheres of activity of the Ministry of Natural Resources and Ecology of the Russian Federation with reliable,

reliable, standardized, metrologically evaluated and legally valid information obtained as a result of the use of a wide range of laboratory methods and equipment (Ozhogina et al. 2017a, b).

2 Methods

The variety and complexity of objects of natural and man-made origin, the widespread use of quantitative methods of mineralogical analysis, the presence of a large number of mineralogical laboratories with different instrumentation and methodological base, personnel composition, determines the need to improve and develop the Quality Management System QA/QC for Mineralogical works UKARM, established at VIMS in the 1980th (Ginsburg et al. 1985) and renovated today. Developed by VIMS the UKARM System coordinates laboratory studies, preparation for testing operations and also monitors the quality of research. Scientific-methodical support of the UKARM functioning carries out the Scientific Council on methods of mineralogical methods of analysis (NSOMMI). The main components of UKARM mineralogical QA/QC system include as follows:

- make the General concept of development of mineralogical service and deliver priority directions of its improvement;
- preparation of proposals for the program and coordination plan of scientific research aimed at solving mineralogical laboratory research methods of testing;
- develop the industry system of standard samples, coordinate their production, registration and use;
- the NSOMMI Scientific Council review, update and elaborate new guidance documents on methods of all kinds of laboratory mineralogical works, the system and means of these documents verification and approval;
- ensuring the functioning of a unified quality management system of laboratory mineralogical work, organization and carrying the interlaboratory comparative tests (ICT), internal and external laboratory analysis control;
- methodological assistance support for the basic and regional laboratories and research centers in certification of mineralogical methods of analysis;
- implementation of complex and unique mineralogical and analytical studies;
- development of operational information system for mineralogical research.

UKARM QA/QC system is based on the requirements of the requirements of ISO/IEC 17025 standard in relation to mineralogical research, starting with the resources as stuff and equipment, ending with metrology and data quality assessment (Fig. 1). The testing laboratory stuff must have the necessary competence to perform at a high level of mineralogical works. UKARM QA/QC includes workshops, advanced training courses for laboratory stuff. For example, every year, since 2011, VIMS arrange annual Seminar "Mineralogical school – Current problems and modern methods". Seminar members discuss the basic state-of-art concepts of mineralogy, prospects of development, mineralogical support of geological exploration, methods of mineralogical analysis, environmental problems, specific features of man-made raw materials study, mineralogy for enrichment of ores, the nature of minerals technological properties and their behavior in geological and technological circulation, mineralogical works metrological and methodological support, etc. Specialists of various affiliation took part in the Seminar.

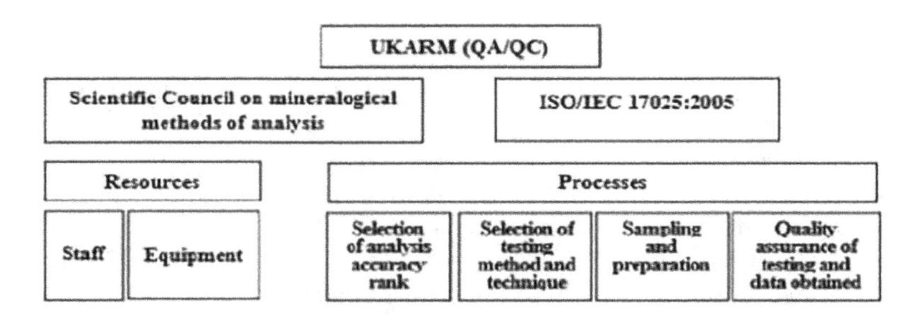

Fig. 1. UKARM QA/QC system structure

Testing laboratories are equipped with the necessary units: measuring instruments, software, standards, reference data, reagents and consumables. Also the methodical base is developed or adapted to technical base of analysis current state, corresponding metrological assessment and its legal approve. The certification and registration of standard samples of phase composition and properties, and artificial mixtures is also a necessary element of QA/QC. The choice of mineral substance testing method and mode is determined by substance peculiarities in composition and structure, the hardware capabilities, the availability of relevant methodological documents and the staff competence. In 2018 the Register on mineralogical studies of natural or technogeneous mineral raw matter includes 3 industry standards, 45 instructions, 31 practical guidelines and 174 recommendations.

Today, the methods of quantitative phase analysis of rocks, ores and man-made substance are especially in demand. UKARM QA/QC includes the classification of quantitative phase analysis laboratory methods, depending on their reproducibility, are divided into six categories: I – particularly accurate quantitative analysis, II – full phase analysis with increased accuracy, III – ordinary quantitative analysis, IV – ordinary analysis with reduced accuracy requirements (express), V – semi-quantitative analysis, VI – qualitative phase analysis. Any testing may/should be characterized by a level of accuracy (Ozhogina et al. 2017a, b).

3 Results and Discussion

The quality of measurements is provided by measuring units' verification and calibration tests; the availability of measurement guidance and their strict observance; internal and external laboratory control. Organization of the control and dispatching service for mineralogical analysis and interlaboratory comparative tests, carrying out the control of phase composition and properties by standard samples, certified mixtures; participation in analysis to certify the standard samples of phase composition and properties, certification of laboratories with mandatory experimental evaluation.

We state the leading role and the one of the most effective forms of external control are *Interlaboratory Comparative testing* (ICT). The last one allow to evaluate the reliability of the results obtained in each individual laboratory, and to obtain information about the real accuracy of measurement techniques in general. In 2016–2018 VIMS conducted a series of ICT on X-ray powder diffraction (XPD) and petrography analysis. The samples for control were artificial mixture of minerals (2), and synthetic corundum powder (1), igneous (1) and sedimentary (1) rocks thing-sections. Each ICT involved from 8 to 12 participants, totally 46 laboratories. The obtained results are considered satisfactory, they can be used by laboratories during accreditation procedures and confirmation of technical competence. The ICT comparative testing provide an opportunity to assess the quality of measurements in different laboratories, to carry out corrective actions to ensure the uniformity of measurements and to show the technical competence of the laboratory.

4 Conclusions

Mineralogical study of any substance should be based on a QA/QC System. Testing laboratory, which performs mineralogical studies supporting geological exploration and technological works in order to meet the requirements of mineralogical work quality management should have the relevant stuff, modern test methods, proper equipment, capabilities and means of verification and calibration, be supplied by industry techniques procedures, guidelines, as well as strictly follow them, and comply with the requirements of a unified quality control system.

References

Ginsburg AI, Vikulova LP, Sidorenko GA (1985) On some typical mistakes at mineralogical investigation (studies) J. In: Zapiski RMO (Proceedings of RMS), vol 3, pp 324–333

JORC Code: 2012 Edition. http://jorc.org/

ISO/IEC 17025:2005: General requirements for the competence of testing and calibration laboratories

Ozhogina EG, Gorbatova EA (2017a) System of mineralogical quality management. J Actual Probl Min 1:3–7

Ozhogina EG, Lebedeva MI, Gorbatova EA (2017b) Interlaboratory comparison tests in the mineralogical works. J. Stand. Samples 2:37–47

14

Absolutely Pure Gold with High Fineness 1000%

Z. Nikiforova(✉)

Diamond and Precious Metal Geology Institute SB RAS, Yakutsk, Russia
znikiforova@yandex.ru

Abstract. It is identified for the first time that, during process of complicated deformation in eolian conditions - mechanical transformation of flaky gold into toroidal form and then into globular-hollow form, gold is cleaned up to absolutely pure metal with fineness 1000‰. Note that, fineness of this eolian gold is higher than fineness of the reference object, shown by the detecting device (JXA-5OA micro-analyzer). In this connection, identified natural process of gold cleaning in eolian conditions can be successfully used in gold metallurgy to obtain absolutely pure gold.

Keywords: Gold · Fineness · Trace elements · Structures · Eolian process

1 Introduction

Native gold in exogenetic conditions, depending on its environment, undergoes gradual changes, in morphology and material composition. It is known that, gold cleaning in exogenetic conditions occurs mainly under chemical and physical-chemical influence in weathering crusts, and also as a result of simple strain in geodynamic conditions. This report will be focused on gold cleaning as a result of impact of mechanogenic processes in eolian conditions.

2 Methods and Approaches

Chemical composition of eolian gold of the Lena-Viluy interfluve (east Siberian platform) was studied by atomic-absorption spectrography (30 objects), spectral quantitative analysis (50 objects), and at JXA-5OA micro-analyzer (30 objects and 200 identifications).

3 Results and Discussion

During mechanical impact of sand grains on gold in eolian conditions, not only the form is transformed, but also inner structure and chemical composition are regularly changed. It is identified that, process of gold cleaning is more intensive in eolian conditions, than in hydrodynamic conditions (Nikiforova 1999).

According to the analysis of eolian gold by atomic-absorption method, increase of gold fineness during transformation of flaky forms into globular-hollow forms from 810 to 970‰ is identified. Flaky gold with scarcely noticeable elevation at the periphery, has fineness range 810–970‰, with average gold fineness 890‰. In toroidal gold, interval of gold fineness fluctuation is identified - 920–970‰, with average index 940‰. Globular-hollow form is characterized by high gold fineness - 960–990‰, with average fineness 970‰.

Spectral quantitative analysis identified that, constant trace elements of flaky forms with characteristic features of eolian transformation are Fe- 0,1; Pb- 0,003; Sb- 0,002 Cu-0,017; Mn- 0,01; Pd- 0,002.; Ni-traces; Hg- traces. And some other. In toroidal gold, a smaller set of trace elements is identified Fe- 0,1; Cu-0,02; Mn- 0,03; Ni- traces; Hg- traces, and in globular-hollow body only these trace elements are identified – Fe- 0,1; Cu-0,05; Mn- 0,001.

Study of fineness of different areas of gold particles (Table 1 and Fig. 1) allowed identifying that, a flake with medium-grained structure (grain C-9a) has fineness from 747 t0 780‰, and its more high-standard shell – from 950 to 988‰. Flake P-138 with partially recrystallized rim has gold fineness 900–970‰, and its central part 814–860‰. Fully recrystallized flake showed maximum gold fineness – 990–1000‰. Fineness of globular-hollow gold within one sample is not just high, but has an absolute value - 1000‰. For example, 17 identifications in the grain 60a found insignificant fluctuations of fineness within interval 992–1000‰, and 13 identifications in the grain 60b showed the highest gold fineness – 1000‰.

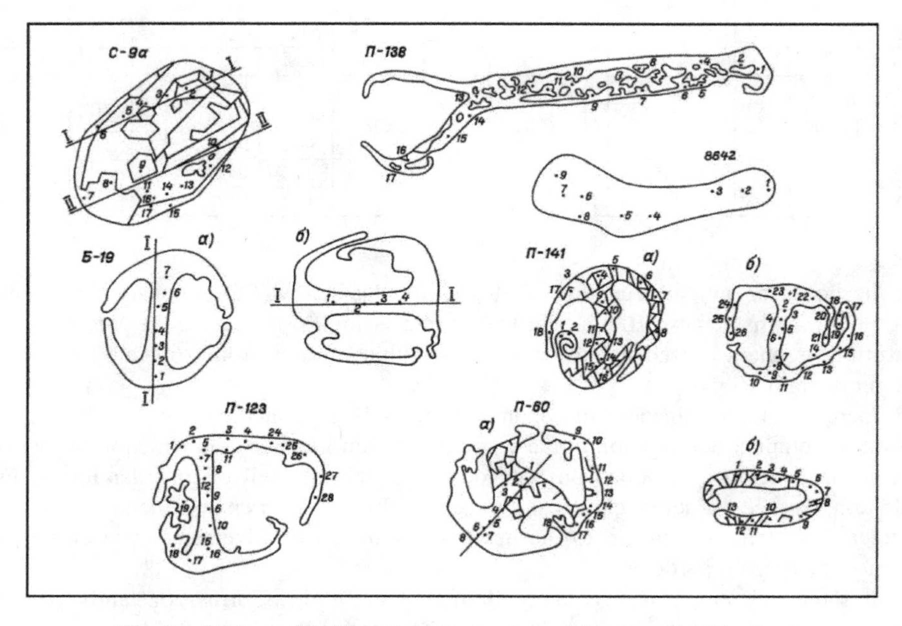

Fig. 1. Section of gold particles and points of fineness identification

Table 1. Fineness of particular areas of gold particles, ‰

Points of fineness measurement	Morphologic type of eolian gold									
	Flaky			Globular recrystallized						
Index of the sample	C-9a	P-138	8642	Б-19a	B-19б	P-141a	P-141б	P-123	P-60a	P-60б
1	759	973	998	903	860	990	997	1000	996	1000
2	747	941	996	912	867	Not identified	992	955	996	1000
3	759	933	998	907	869		983	997	999	1000
4	756	Not identified	984	904	919	990	991	999	1000	1000
5	757	908	994	912		988	985	996	999	1000
6	755	920	997	902		996	996	995	997	1000
7	758	860	996	901		980	987	1000	992	1000
8	769	892	997			999	984	938	992	1000
9	766	978	1000			980	970	930	999	1000
10	953	814				989	995	993	994	1000
11	783	896				994	985	997	1000	1000
12	979	806				993	995	995	1000	1000
13	986	956				990	987	not identified.	1000	1000
14	970	937				920	995		1000	
15	974	973				932	1000	903	1000	
16	988	974				994	988	912	1000	
17	970	900				998	987	910	1000	
18						987	1000	1000	1000	
19							998	999		
20							984	Not identified		
21							991			
22							1000			
23							983			
24							996	1000		
25							998	1000		
26							999	1000		
27								1000		
28								1000		

It should be emphasized that, shell of globular-hollow forms constantly shows absolutely high fineness 1000‰. Just in several samples, primary gold preserved in partition, is characterized by lower fineness within a range from 860 to 919‰, for example grain 19b.

Increase of gold fineness in eolian conditions is explained by the fact that, as a result of complex deformation of flaky gold, very thin films of gold (fraction of mcm) are formed, being overlapped on each other, generate a shell of globular forms. In addition, surface for active chemical interaction of metal with environmental elements is increased, that contributed to the maximum removal of silver and trace elements from primary gold particles.

It should be emphasized that, shell of globular-hollow forms constantly shows absolutely high fineness 1000‰. Just in several samples, primary gold preserved in partition, is characterized by lower fineness within a range from 860 to 919‰, for example grain 19b.

Increase of gold fineness in eolian conditions is explained by the fact that, as a result of complex deformation of flaky gold, very thin films of gold (fraction of mcm) are formed, being overlapped on each other, generate a shell of globular forms. In addition, surface for active chemical interaction of metal with environmental elements is increased, that contributed to the maximum removal of silver and trace elements from primary gold particles.

The process of mechanically gold cleaning was proved by (Lechtman 1979) experimentally. According to the results of his experiments, clean layer of gold appeared after multiple alteration of forging of gold and copper alloy, with its processing in weak ammonia solution, where initial gold content did not exceed 12%.

4 Conclusions

It is identified for the first time, that during complex deformation, when flaky gold is mechanically transformed into toroidal form and then into globular-hollow form, metal is cleaned to absolutely pure gold with fineness 1000‰. Identified natural process of gold cleaning in eolian conditionsв can be successfully used in gold metallurgy to obtain absolutely pure gold.

Acknowledgements. The work is implemented within scientific-research projects of Diamond and Precious Metal Geology Institute, Russian Academy of Sciences, project № 0381-2016-0004.

References

Lechtman H (1979) A pre-Columbian technique for electrochemical replacement plating of gold and silver on copper objects. J Metals 31(12):154–160
Nikiforova ZS (1999) Typomorphic features of eolian gold. ZVMO. N5, pp 79–83

15

Application of Fluoride Technology for Processing of Off-Grade Aluminum Raw Materials

I. Burtsev, I. Perovskiy(✉), and D. Kuzmin

Institute of Geology named after Academician N.P. Yushkin Komi Science Center of the Ural Branch of the Russian Academy of Sciences, Syktyvkar, Russia
igor-perovskij@yandex.ru

Abstract. We used the process of hydrofluoride desilication for off-grade aluminum raw materials widely distributed in the Komi Republic. The optimal ratios of fluorinating agent (NH_4HF_2) to the target component (SiO_2) were determined by the method of differential thermal analysis. At a fluorination temperature 300 °C and timing 30 min, we obtained concentrates containing up to 80% Al_2O_3 and up to 10% SiO_2.

Keywords: Bauxite · Aluminum raw materials · Desilication · Fluorination · Ammonium hydrofluoride

1 Introduction

One of the most important problems of Russian aluminum industry is the deficit of high-quality alumina raw materials, forcing to import it from other countries.

A major source of alumina raw materials is the Vezhayu-Vorykvinskoe and Verkhneshchugorskoe bauxite deposits, which are part of the Vorykvinskaya group of deposits in Middle Timan and developed by the United Company RUSAL. Mined bauxites are predominantly consumed for alumina production by the Bayer method.

The share of low-quality sintered bauxites is significant and amounts from 5 to 55% of balance reserves for Vezhayu-Vorykvinskoe and Verkhneshchugorskoe deposits. A sharp reduction in consumption of sintered bauxites makes an alternative processing a rather actual question. In addition, the prospected reserves contain a significant amount of off-balance ores, off-grade in quality: high-silica low-modulus (better M_{Si} 2.6–3.8) bauxites and associated high-aluminous allites.

Gravitational, flotation, magnetic, chemical, and other methods have been proposed to enrich the bauxites, but their use for enriching Middle Timan bauxite is limited due to a fine size of the minerals, a large proportion of amorphous phases, and a small efficiency of used processes in general (Burtsev et al. 2016).

Fluoride technologies are one of the most promising ways to process mineral raw materials (Dyachenko and Kraidenko 2007; Medkov et al. 2011). The purpose of the research – to assess the prospects for the use of fluorination to desilicate off-grade aluminum raw materials (bauxite, allite, kaolinite clays), which are widespread in the Komi Republic.

2 Methods and Approaches

The objects of research - representative samples of bauxite, allite, kaolinite clays from the Vezhayu-Vorykvinskoe (V-V), Verkhneshchugorskoe (Vsch) deposits and occurrences of the Izhma area (IA). Samples were grinded in a disk eraser (ID-200) and separated by size classes 0.125–0.250 and 0.25–0.5 mm. Analytical work was carried out by the equipment of the Center for Collective Use GeoScience of the Institute of Geology Komi SC UB RAS using XRD, XRF, DTA. Fluorination was carried out in a tube furnace equipped with a gas-extraction system. Ammonium hydrodifluoride (NH_4HF_2) was used as a fluorinating component, which is an ecologically safe matter under standard conditions.

3 Results and Discussion

X-ray fluorescence analysis showed that the granulometric differentiation did not lead to a change in the chemical composition and silicon module of the samples (Table 1). We diagnosed reflexes of kaolinite with hematite admixture on diffraction patterns of samples No. 1, 2, and 3, and boehmite, kaolinite with anatase admixture on the patterns of sample No. 4.

Table 1. Chemical composition of off-grade bauxite ores

No.	*Class, mm*	*Mass fraction, %*						M_{Si}
		Al_2O_3	SiO_2	Fe_2O_3	TiO_2	MgO	K_2O	
1	0.25–0.5	43,82	50,83	2,93	1,66	0,31	0,10	0.86
IA	0.125–0.25	44,08	50,76	2,70	1,66	0,31	0,10	0.86
2	0.25–0.5	43,32	48,40	4,96	1,59	0,67	0,67	0.88
Vsch	0.125–0.25	43,63	48,67	3,92	1,78	0,79	0,71	0.90
3	0.25–0.5	35,94	39,83	19,84	1,37	1,76	0,68	0.90
V-V	0.125–0.25	36,01	39,99	19,75	1,27	1,77	0,65	0.90
4	0.25–0.5	59,85	23,91	12,98	1,12	1,03	0,61	2.50
Vsch	0.125–0.25	59,37	23,78	13,53	1,15	1,04	0,62	2.50

Admixture components Na_2O, CaO, MnO, P_2O_5 no more than 0.8%.

Thermodynamic calculations of fluorination of kaolinite NH_4HF_2 are given elsewhere (Rimkevich et al. 2016). However, the theoretical equation turns out to be complicated in stoichiometric terms, which leads to the overconsumption of the fluorination component. Our works on fluorination of titanium ores of the Yarega deposit in the Komi Republic showed the efficiency of applying the ratio calculation to the target component—silicon oxide (Perovskiy and Ignat'ev 2013; Perovskiy and Burtsev 2016). SiO_2 fluorination can be described by the following equation:

$$SiO_2 + 3NH_4HF_2 = (NH_4)_2SiF_6 + 2H_2O + NH_3 \uparrow$$

We studied the process of fluorination of off-grade bauxites at molar ratios SiO_2: NH_4HF_2 equal to 1: 1 and 1: 1.5 with the help of DTA. We established that ratio 1:1 was preferred. Excessive ammonium fluoride (1:1.5) did not result in a positive effect, but was accompanied by passivation of fluorination reactions and formation of a larger volume of gaseous products.

Taking into account DTA data, the temperature regime of fluorination consisted of sintering the sample with NH_4HF_2 at temperature 200 °C (30 min.) with subsequent sublimation of resulting salt $(NH_4)_2SiF_6$ at 300 °C (30 min.). Upon completion of the fluorination, we carried out water leaching, which allowed transferring the undecomposed fluoroammonium salts into the solution.

The results of fluorination, given in Table 2, show that SiO_2 content is significantly reduced in the samples. At the same time, the size of material does not affect the effectiveness of desilication.

Table 2. Chemical composition of samples after fluorination

No.	*Class, mm*	*Mass fraction, %*					
		Al_2O_3	SiO_2	Fe_2O_3	TiO_2	MgO	K_2O
1	0.25–0.5	80.77	7.75	6.75	3.60	0.33	0.25
IA	0.125–0.25	81.17	7.73	6.78	3.67	0.32	0.23
2	0.25–0.5	73.94	8.98	9.85	3.92	1.03	1.66
Vsch	0.125–0.25	71.81	11.24	10.07	3.67	1.00	1.58
3	0.25–0.5	55.07	7.28	30.20	2.38	2.72	1.40
V-V	0.125–0.25	52.87	9.17	30.61	2.29	2.74	1.36
4	0.25–0.5	68.00	9.99	17.55	1.45	1.46	1.01
Vsch	0.125–0.25	67.31	9.67	18.52	1.48	1.44	1.02

Admixture components Na_2O, CaO, MnO, P_2O_5 no more than 0.9%.

4 Conclusions

The process of hydrofluoride desilication has been applied to off-grade aluminum raw materials widespread in the Komi Republic. The optimal ratio of NH_4HF_2: SiO_2 equal to 1:1 was determined by DTA method. At the fluorination temperature 300 °C and timing 30 min, concentrates with Al_2O_3 content more than 70–80%, and SiO_2 - less than 10% were produced. This technology is an alternative to the acid and alkaline processing of high-siliceous aluminum raw materials. The use of fluoride technology allows not only to improve the quality of ores because of desilication, but also to obtain products that can be directly processed by electrothermal methods to silumin (Lepezin et al. 2014), ferroalloys (Bukin and Seregin 2014) and other products with high added value.

Acknowledgements. The work was carried out with the financial support of the State task "Scientific basis for effective subsoil use, development and exploration of mineral resource base, development and implementation of innovating technologies and economic zoning of the Timan-Nothern Ural region" No. AAAA- A17-117121270037-4.

References

Bukin AV, Seregin AN (2014) Development of technology for smelting ferrosilicoaluminium from sub-standart bauxite and aluminous waste of metallurgy and electrical power energetic. Probl Ferr Metall Mater Sci (2):31–36

Burtsev IN, Kotova OB, Kuzmin DV, Mashin DO et al (2016) Prototypes of new technologies for development of the mineral raw materials complex of the timan-north urals region. In: Proceedings of the Komi science centre of the Ural division of the Russian academy of sciences, vol 27, no 3, pp 79–88

Dyachenko AN, Kraidenko RI (2007) Separation of silicon-iron-copper-nickel concentrate by fluorammonium method into individual oxides. Bull Tomsk Polytech Univ 311(3):35–38

Lepezin GG, An'shakov AS, Faleev VA, Avvakumov EG, Vinokurova OB (2014) Plasma-chemical method of producing silumin and aluminium from the minerals of the sillimanite group. Doklady Chem 456(2):110–113

Medkov MA, Krysenko GF, Epov DG (2011) Ammonium bifluoride – the perspective reagent for complex processing of mineral raw materials. Vestn Far East Branch Russ Acad Sci. 159 (5):60–65

Perovskiy IA, Burtsev IN (2016) Influence of mechanical activation of leucoxene on efficiency of its processing by fluoride method. Perspektivnye Materialy (2):66–73

Perovskiy IA, Ignat'ev GV (2013) Ammonium fluoride method of desilication of leucoxene concentrate of Yarega deposit. In: Predictive estimate of technological properties of minerals by applied mineralogy methods. Proceedings of 7th Russian seminar of process mineralogy, Karelian Scientific Centre RAS, Petrozavodsk, pp 110–116

Rimkevich VS, Pushkin AA, Girenko IV, Leontyev MA (2016) Perspectives of complex processing the kaolin concentrates by hydrochemical method. Izv Samara Sci Cent Russ Acad Sci 18(2):186–190

16

Mineral-Geochemical Criteria to Gold and Silver Recovery for Geometallurgical Sampling Campaign on Primorskoe Gold-Silver Deposit

I. Anisimov(✉), A. Sagitova, O. Troshina, and I. Agapov

Technology Research Department, Polymetal Engineering JSC, St-Petersburg, Russia
anisimovis@polymetal.ru

Abstract. Ore variability study at Primorskoe gold-silver deposit demonstrated wide variety of mineral composition and gold and silver recoveries with cyanadation. The ore consisted of quartz-feldspar veins, quartz-rhodonite, quartz-"pyrolusite", quartz-epidote-garnet and quartz-Mn-silicates/hydroxides. Todorokite, birnessite, rancieite were the most common among the last ones. Statistical analyses of chemical and mineral composition, parameters of cyanidation tests showed occurrence of three main ore types – feldspatic, manganese silicate and oxide. Gold recovery effected by locking in Fe-oxides. The highest silver recovery strongly correlated with feldspathic cluster and whiter sample color reflecting Ag mineral forms: acanthite and electrum. Ore with silicate Mn showed good recoveries of acanthite, electrum and iodargirite associated with Mn silicates. Main silver losses were connected with Mn-oxides content and dark ore coloration where Ag chemically bound in Mn-oxides. Sr and Ba content along with sample color were indications that could be used as a proxy for recovery in geometallurgical mapping and ore-sorting.

Keywords: Silver · Predictive recovery · Cyanidation · Manganese mineralization

1 Introduction

Primorskoe Au-Ag deposit is located in Omsukchan district of Magadan oblast, Russia. It formed by hydrothermal veins of various composition: Qu-Fsp-Chl with Ep, Qu-Rdn-MnO_x, Qu-MnO_x and Qu-Gar ± Wol. Noble metal mineralization represented with acantite, iodargirite, aurorite, jalpaite, pyrargyrite, electrum, kustelite and native silver and gold. Abundance of specific manganese oxides effected gold and silver extraction by cyanidation. The mineral composition study was aimed to define causes of possible gold and silver losses with cyanidation tails and possible ways to recover refractory silver.

2 Methods and Approaches

72 individual and 57 composite geometallurgical samples of Primorskoe deposit were studied for chemical and mineral composition and cyanidation tests were performed. Ag varied from 176 to 36450 ppm in individual samples and gold – from 0.01 to 113 ppm. Ag content range in the composite samples was 49–1509 ppm averaging 606 ppm and Au – 0.04–9.92 ppm averaging 1.56 ppm.

Chemical composition was studied at ALS labs, Moscow. Au content was assayed by fire assay with atomic absorption finish, Ag and other 35 elements analyzed by ICP-AES after four acid digestion.

Quantitative X-ray powder diffraction by Rietveld refinement (QXRD) was carried out using Bruker D8 Advanced diffractometer with Linxeye XE detector at Technology Research Department, Polymetal Engineering.

Statistical analysis was done on filtered data with Aitchison transformation using PCA, regression analyses and Pearson correlations in Cytoscape software.

3 Results and Discussion

Mineral composition of the sample significantly varied in quartz (up to 94%), feldspar (up to 63%), micas (illite and muscovite – up to 25%). Epidote and piemontite formed up to 14.5%, grossular and calderite – up to 8.9%. Bustamite (up to 22.8%), rhodonite (up to 14.6%) and wollastonite (up to 16.7%) were observed in some samples. Apatite-magnetite-titanite-vermiculite association was common in few samples. Manganese oxides contents ran up to 34.4%.

Mn-mineralization presented in 18 minerals: silicates (piemontite, calderite, bustamite, rhodonite), carbonate (rhodochrosite), sulfide (alabandite), oxi-hydroxides. Todorokite, birnessite (7Å, 14-Å and amorphous), rancieite well spread among the last ones. Manganite, pyrolusite, ramsdellite, jacobsite, bixbyite, pyrochroite, chalcophanite, coronadite, aurorite were less common.

Data population for multivariate statistical analyses included 103 following parameters of chemical and mineral composition, color (RGB, brightness – BRT, and darkness - DRN), material fineness (γ + 100), silicate and oxide Mn ratio (Mnsil/Mnox), 14 parameters of cyanidation conditions (NaCN concentration in the pregnant solution – NaCNps; NaCNc and CaO consumption; final test pH – pHf; Au and Ag contents in pregnant solution and cake – AuPS, AgPS, AuK, AgK; Au and Ag recovery – εAu, εAg and losses – $-\varepsilon$Au, $-\varepsilon$Ag (Fig. 1). Pearson correlations revealed occurrence of two main geochemical and mineral clusters – whiter feldspathic (rock-forming Ab-Olg-Chl) and darker manganese. Mgt-Tit-Ap-Vrm sub-cluster occurred in the feldspathic one showing accessory syenite mineralization. "Skarn" association (wollastonite, diopside, andradite, calcite, bustamite, rhodonite) and quartz vein (Qu, SiO_2, Cr and Iron – grinding contaminants) were distinguished. Ag recovery and NaCN content in pregnant solution tied with color cluster.

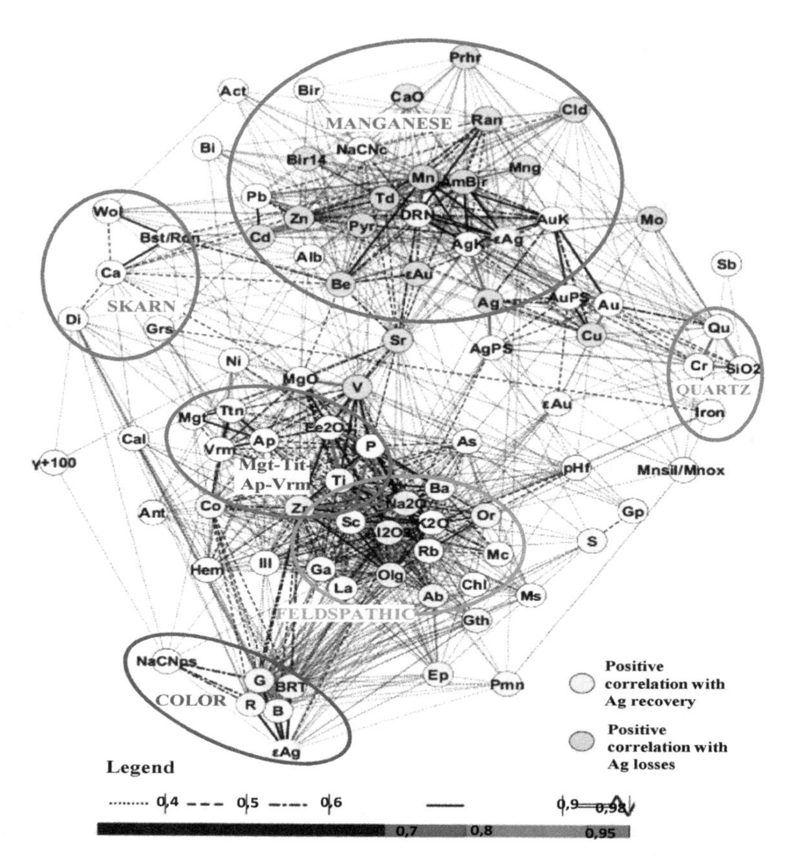

Fig. 1. Correlations between 88 transformed parameters of individual and composite samples of geometallurgical sampling campaign of Primorskoe deposit and proposed mineral association clusters. Legend shows Pearson correlation coefficient values

Reagents consumptions (CaO, NaCN) correlated with Mn-cluster, proving that ion exchange might take place during leaching. Zn, Cd, Pb tended to Mn-cluster without connection to a specific Mn-oxide. Sr related to both main clusters equally reflecting isomorphic distribution in feldspathic gangue as well as in MnO_x.

Ag recovery demonstrated positive correlation with feldspathic cluster and sample white color (BRT). Ag losses had strong connection with MnO_x. Au recovery tied together with color and quartz, Au losses correlated with MnO_x and locking.

PCA analysis showed same regularities. 6 principal components explained 54.78% of the total variance and described mineral composition (Fig. 2), Mn- and accessory mineralization, noble metal contents and recovery. PCA highlighted strong connection between Ag loses, todorokite and amorphous birnessite.

Regression Eq. (1) showed strong relation (R^2 = 0.66) between silver recovery, color (BRT) and some elements contents (in ppm). Variables listed in the order of their significance:

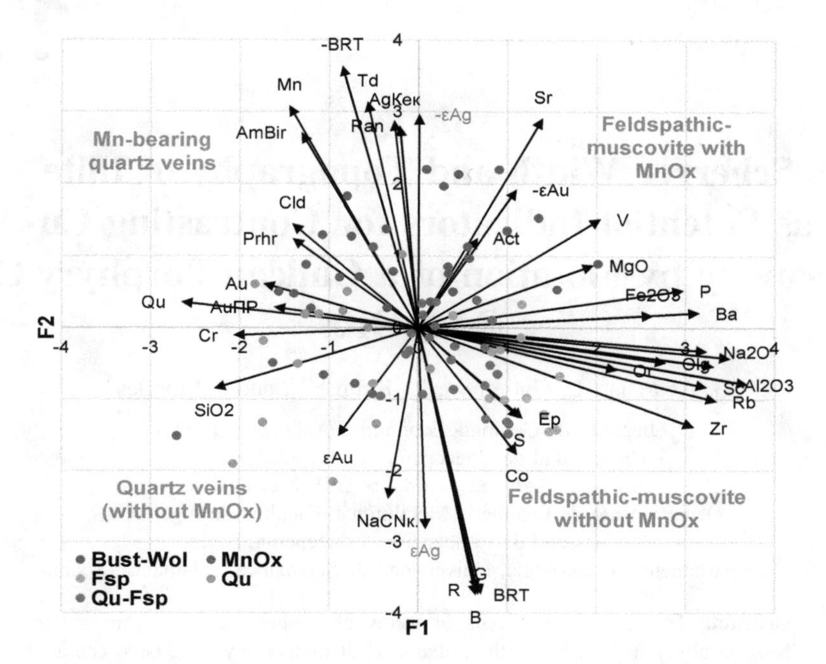

Fig. 2. Factor loadings and factor scores for factors 1 and 2 with their interpretation

$$\varepsilon Ag = 18.24 + 0.47BRT + 1.39Al2O3(wt\%) - 0.12V + 0.05Sr - 3.45Sc - 0.01Ag - 1.21Mn(wt\%) + 0.45Ca(wt\%) + 0.003Cu - 0.13Rb + 0.01S \quad (1)$$

4 Conclusions

Individual and composite samples of Primorskoe Au-Ag deposit demonstrated variety of mineral composition from high silica and feldspar to Mn-skarn association. Mn mineralization presented in silicates, oxides and hydroxides, carbonate, sulfide had great impact on Ag recovery. Wide range of vein gangue, ore and noble metal mineral associations were similar to ones characteristic for Dukat deposit, the largest silver deposit in Russia.

Ag mineral forms affected recovery rate from highest presented by acanthite, electrum in feldspatic association to the lowest locked in Mn-oxides, mainly in todorokite and birnessite. Ore sorting by color and element content can be used.

Acknowledgements. The authors are grateful to Sergey Kubyshkin for performance of cyanidation tests and Boris Milman for scientific advices.

17

Scherrer Width and Topography of Illite as Potential Indicators for Contrasting Cu-Recovery by Flotation of a Chilean Porphyry Cu (Mo) Ore

G. Abarzúa[1], L. Gutiérrez[2], U. Kelm[1(✉)], and J. Morales[3]

[1] Instituto de Geología Económica Aplicada (GEA), Universidad de Concepción, Concepción, Chile
ukelm@udec.cl, ued.kelm@gmail.com

[2] Departamento de Ingeniería Metalúrgica, Facultad de Ingeniería, Universidad de Concepción, Concepción, Chile

[3] Departamento de Geología, Universidad de Salamanca, Salamanca, Spain

Abstract. Two contrasting feeds in terms of copper recovery from a Cu (Mo) porphyry deposit but with similar overall mineralogy have been characterized by X-ray diffraction for their <1 μm fraction illite crystallinity, Scherrer width and by atomic force microscopy for surface roughness. The unfavorable feed displayed slightly higher crystallinity, larger Scherrer width and surface roughness factors, than the feed with good Cu recovery. As Scherrer width is an easy and cheaply to determine parameter it is suggested as complementary information to particle size distribution analyses when dealing with feeds where illite may affect pulp viscosity or gangue adhesion to bubbles during flotation.

Keywords: Illite · Scherrer width · Surface roughness · Flotation · Sericite · Atomic force microscopy

1 Introduction

For over half a century, illite crystallinity has been used as an indicator of mineral maturity in metasediments between the transition of diagenesis to very low temperature metamorphism and the incipient low-grade metamorphism or epizone (Frey 1999). Illite crystallinity (later Kübler Index) measured at full width half medium height (FWHM) of the basal XRD-reflection is also an indirect indicator for the size of the jointly diffracting illite sheets, also known as Scherrer width, which has been directly visualized with the widespread availability of Transmission Electron Microscopy (Frey 1999). Superimposition of metamorphic and hydrothermal alteration processes, paired with time consuming analytical routines, has limited the application to alteration halos of ore deposits; an exception is the study by Beaufort et al. (2005) on the East Alligator River Uranium deposit in the Northern Territory, Australia, due to its abundance of chlorite and illite gangue. Sericitic alteration (muscovite/illite) also represents widespread gangue for Andean Cu (Mo) porphyry ore deposits, but systematic studies of phyllosilicate crystallinity or Scherrer width within ore deposit areas are not available.

Cheng and Peng (2018) suggest negative effects for low crystallinity kaolinite rich ore, much of the work being based on artificial ore-gangue mixtures. However Jorjani et al. (2011) single out illite and vermiculate as key gangue affecting flotation for the Iranian Sarcheshmeh porphyry copper deposit. The present exploratory study has been sparked by the effort to develop a formula for blending ore based on mineralogical-chemical parameters of a giant porphyry copper deposit in Chile. Ore with similar sericitic alteration and chalcopyrite dominated ore phases, were identified by the concentrator operation as favorable (F) and unfavorable (UF) floating feed, the latter entering the concentrator only as blend. Given this overall mineralogical similarity, it was decided to characterize the contrasting feeds based on their illite crystallinity – Scherrer width and concomitant surface roughness.

2 Methods and Approaches

Triplicate samples of favorable (F) and unfavorable (UF) floating ore were prepared for X-ray diffraction (XRD) analysis of the clay size fraction ($>0.45 <1$ μm) based on the recommendations of Moore and Reynolds (1997). XRD measurements were carried out on a Bruker D4 diffractometer operated with Ni-filtered Cu-radiation. Illite crystallinity and Scherrer width were determined on the 001 basal reflection following Lorentzian adjustment using the Origin 8.5 program. Atomic Force Microscope (AFM) measurements of topography were carried out with an AIST-NT equipment in contact mode on a 5×5 μm surface. WSxM5.0 software was used for calculating Ra (arithmetic average) and Rrms (root mean square) roughness factors (Horcas et al. 2007; Erinosho et al. 2018).

3 Results and Discussion

Illite crystallinity values of both samples (F: 0.15 $\Delta° 2\Theta$, UF: 0.12 $\Delta° 2\Theta$) correspond to epizone values, and for the unfavorable feed are at the sensitivity limit of this method (Frey 1999). Though values for the favorable feed are marginally lower, nevertheless this difference is expressed in an increased Scherrer width or crystallite size (F: 48.8 nm, UF: 62.6 nm) for the unfavorable feed.

Topographic images of sample surfaces show different roughness, being the favorable feed (F) the smoother. Roughness factors were calculated for different surfaces scanned by AFM. As a mean, 10 sample surfaces were measured and statistically compared, giving values of Ra = 57 ± 20 and Rrms = 66 ± 23 for the favorable feed and Ra = 68 ± 18 and Rrms = 86 ± 19 for the unfavourable feed sample. Undoubtedly, different scales of images analysed imply changes in the surface parameters. To avoid this, images with the same size have been compared. As it can be observed in the Fig. 1, roughness parameters confirm the XRD results.

Despite the difficulties of differentiating between illite generations in rocks with sericitic overprint in porphyry (and other) ore deposits, this simple XRD measurement permits concomitant calculation of coherently diffracting particle sizes or Scherrer width for a given geometallurgical unit. The correspondence observed for this

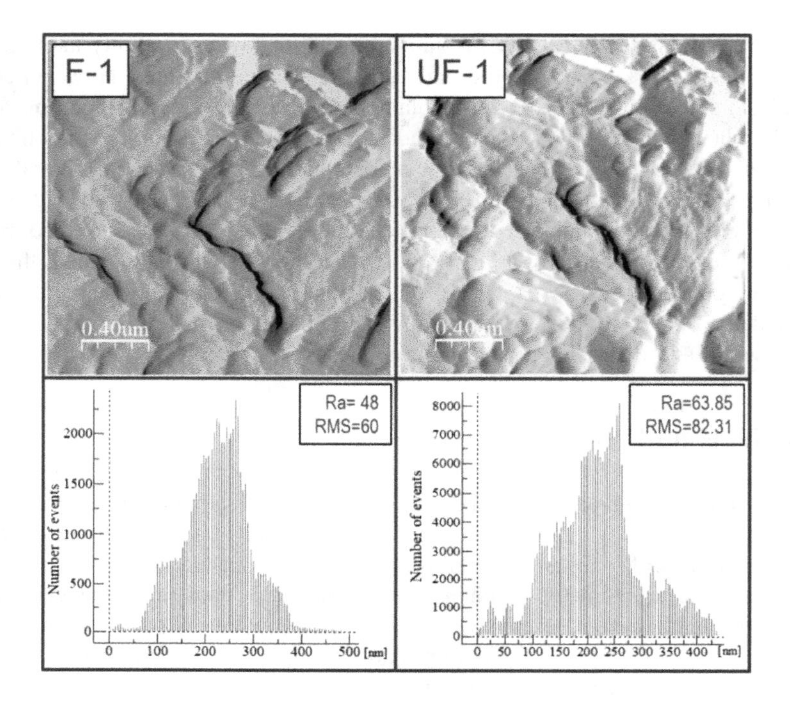

Fig. 1. Topographic aspects of F-1 sample (left) and UF-1 sample (right)

exploratory study between increased Scherrer width and higher AFM measured surface roughness, recommends the XRD based value as an easily available tool to assess difference in flotation for samples where traditional methods like optical microscopy, automated mineralogy, and semi-quantitative whole rock mineralogy do not reveal obvious mineralogical differences between feeds with contrasting flotation behavior. Farrokhpay and Ndlovu (2013) discussed the effect of clay particle size on pulp rheology; here the X-ray coherent Scherrer width is suggested as a complementary indicator to the particle size distribution measurements by laser diffraction in the clay size range. The scale of surface roughness as a factor impacting on particle adhesion to bubbles has been studied by Nikolaev (2016). However, direct AFM measurements are still no routine procedures to define geometallurgical units within an ore deposit, whereas XRD information can be generated faster and in a more standardized fashion.

4 Conclusions

Illite crystallinity, Scherrer width and AFM-determined surface roughness have been determined for two flotation-feed of a Cu (Mo) porphyry copper deposit with contrasting Cu-recovery. Samples did not display any mineralogical difference allowing a straightforward explanation of this difference. For this exploratory study case, Scherrer width is considered an easy to obtain parameter that points to differences in surface roughness of illite particles in the <1 μm fraction and thus may influence pulp viscosity and/or particle adhesion to bubbles during flotation.

Acknowledgements. Dr Manuel Melendrez, Departamento de Ingeniería de Materiales, Universidad de Concepción is thanked for access to the AFM equipment.

References

Beaufort D, Patrier P, Laverret E (2005) Clay alteration associated with proterozoic unconformity-type uranium deposit in the East Alligator Rivers uranium fields, Northern Territory, Australia. Econ Geol 100:515–536

Chen X, Peng Y (2018) Managing clay minerals in froth flotation. A critical review. Miner Process Extr Metall Rev 39(5):289–307

Erinosho MF, Akinlabi ET, Johnson OT (2018) Characterization of surface roughness of laser deposited titanium alloy and copper using AFM. Appl Surf Sci 435:393–397

Farrokhpay S, Ndlovu B (2013) Effect of phyllosilicate minerals on the rheology, colloidal and flotation behaviour of chalcopyrite mineral. In: Australasian conference on chemical engineering, Chemeca 2013, Challenging Tomorrow, p 733

Frey M (1999) Very low-grade metamorphism of clastic sedimentary rocks: in Low Temperature Metamorphism. Blackie and Sons, Glasgow, pp 9–58

Horcas I, Fernández R, Gomez-Rodriguez JM, Colchero JWSX, Gómez-Herrero JWSXM, Baro AM (2007) WSXM: a software for scanning probe microscopy and a tool for nanotechnology. Rev Sci Instrum 78:013705

Jorjani E, Barkhordari HR, Khorami MT, Fazeli A (2011) Effects of aluminosilicate minerals on copper–molybdenum flotation from Sarcheshmeh porphyry ores. Miner Eng 24(8):754–759

Moore DM, Reynolds RC Jr (1997) X-ray diffraction and the identification and analysis of clay minerals. Oxford University Press, Oxford

Nikolaev A (2016) Flotation kinetic model with respect to particle heterogeneity and roughness. Int J Miner Process 155:74–82

18

Ore Mineralogy of High Sulfidation Çorak-Taç Epimesothermal Gold Deposit

K. Diarra[1], E. Sangu[1], and E. Çiftçi[2(✉)]

[1] Department of Geological Engineering, Faculty of Engineering, KOU, 41380 Kocaeli, Turkey

[2] Department of Geological Engineering, Faculty of Mines, ITU, 33469 Maslak, Istanbul, Turkey

eciftci@itu.edu.tr

Abstract. Çorak and Taç, two nearby mineralizations, are located in the eastern black sea region, which is one of the most productive metallogenic belts of Turkey. It is characterized by a great number of Kuroko-type volcanogenic massive sulfide deposits as well as vein-type polymetallic deposits, porphyry and epithermal precious metal deposits. Subject neighboring deposits are hosted within the voluminous Cretaceous-Eocene granitoids and interbedded volcanic rocks and carbonates. Mineralogy of altered host rocks include quartz veins, carbonates, sericite, chlorite, chalcedony, and disseminated sulfides - mainly pyrite, sphalerite, galena, and chalcopyrite. The main texture encountered in the host rocks is hyalo-porphyry. Due to hydrothermal alterations primary minerals are mostly altered in which the ferromagnesian minerals are chloritized and calcified, while feldspars are altered into sericite, calcite, and albite. Silicification and argillic alteration (medium, moderate, high) are widely spread however; XRD analysis carried on 33 core samples from Çorak has also revealed local propylitic alteration, limonitization and hematitization as well. The minerals assemblages that accompanied the different alterations include jarosite and alunite suggesting high sulfidation hydrothermal mineralization. Through the ore microscopic studies, pyrite, chalcopyrite galena, sphalerite, and a lesser amount of sulfosalts (tennantite-tetrahedrite and pyrargyrite) were determined. Quartz and calcite account for the main gangue minerals. While the Taç mineralization is pyrite, chalcopyrite and sphalerite dominating, the Çorak mineralization contains relatively less chalcopyrite and galena becomes prevalent with sphalerite. Gold in both sites may reach up to 10 ppm, on average 3 ppm. Silver occurrence is insignificant.

Keywords: Yusufeli · Artvin · Turkey · High sulfidation · Gold · Epimesotherma

1 Introduction

The study area is located in Yusufeli area in the eastern Pontides' northeastern most tip within a larger metal rich tectonic corridor stretching from southern Georgia and northern Armenia to Bulgaria and Romania and it is a host to numerous volcanogenic massive sulfide (VMS) and vein-type deposits dominantly of Late Cretaceous age.

Taç and Çorak deposits occur in an area about 10 km × 3 km in size, oriented NW-SE and SW-NW, respectively. It is limited to the north by series of outcrop of large unaltered basaltic basement with some andesite enclave. They are covered by more recent tuffs, sandstone and old alluvial fan. The south is more dominated by andesites and tuff outcrops together with relatively recent sedimentary covers.

Mineralization and alteration are represented by extensive strata-bound argillic alteration cut by gold and base metal bearing quartz veins within a massive pyroclastic volcanic host-rock below a bedded volcanogenic succession, with additional receptive massive volcanic units occurring higher up in the sedimentary succession. Porosity is believed to be the most likely control on the stratiform nature of alteration. However, proximity to major structures along or close to the structural axis appears to be an important influence on the presence of alteration suggesting structural control on mineralization.

Some typical structures are associated with the Cu-Pb-Zn mineralization: The most obvious one is the hydrothermal alteration with its colored intense argillic and propylitic alterations. They are always associated with a local fault. Afterward, the large intrusive rocks are crosscutting the volcanic, the pyroclastic and the volcano sedimentary rocks.

In Çorak SW-NE section view, the high grade of base metal representing the mineralization is localized in volcano sedimentary rock (tuff) as well as in andesite and feldspar porphyry andesitic, the three sets are separated by set of faults.

2 Methods and Approaches

For mineralogical and ore microscopy studies, 35 thin and polish sections from core drill and outcrop, have been prepared and scrutinized under reflect and transmitted light microscope of the laboratories of Kocaeli university (KOÜ) and Istanbul technical university (ITU) in Turkey. 6 samples have been taken from alteration zone have been analysis in order to compute their CIA: $([Al_2O_3/(Al_2O_3 + CaO + K_2O + Na_2O)]$ 100) and ICV: (ICV: $CaO + K_2O + Na_2O + Fe_2O_3$ (t) $+ MgO + MnO + TiO_2)/Al_2O_3)$. The target was to determine their depositional environment as well as the alteration styles. 33 samples from different alteration zones from Çorak have been submitted to the lithological and XRD analysis using Rietveld methods.

15 polished thin sections have been subjected to Cathodoluminescence (CL) microscopy study to describe and to distinguish gangue minerals and minerals generations. Samples containing mostly sphalerite, quartz and carbonates were chosen, since they are able to produce CL emissions. The CL study was performed at the ITU Advanced Microscope lab using an optical cathodoluminescence system (CITL MK5 system).

3 Results and Discussion

The andesite and andesitic tuffs are the main lithologies hosting the mineralization in the study area. The mineralogy of the altered hostrock enclosed quartz veins, carbonate, sericite, chlorite, and chalcedony and disseminated sulfides. Their details microscopic studies have revealed the following common textures: trachytic; hyalo microlithic porphyry; vesicular microlithic porphyry; hyalo micro granular porphyry and quasi-oolithic. The hydrothermal alteration manifest itself in the form of silicification, epidotization, chloritization, seritization, and finally a calcification as a last stage. Colored argillic alteration is widespread in the area along with silicic and propylitic alteration. Drill holes intercept sulfide ore zones at different depth. The sulfide mineralizations may range from 50 cm to 1 m in thickness. Weakly mineralized zones of 1 to 2% with disseminated sulfides are intercalated between the massive or semi massive sulfide lenses. The altered interval shows 3 types of minerals assemblages: Quartz + sericite + calcite, quartz + epidote and quartz + chlorite + calcite: (I) **quartz + chlorite alteration**: In this facies, the ferromagnesian (Amphibole) are completely chloritized and calcified. Chlorites are filling cavity as well. They wrap also the fewer amount of chalcedony present in the groundmass. The opaque minerals represent less than 1–2% of the groundmass. The texture is vesicular hyalo-microlithic porphyry; (II) **quartz + epidote alteration:** In this facies, the light colored plagioclases are turned into epidote. The calcite as a last phase is filling fracture. (propylitic alteration)**;** **(III) quartz + sericite alteration:** The sericite is not only replacing the ferromagnesian and the plagioclase but also the matrix. Plagioclases turned into sericite. Euhedral quartz is present in the groundmass. The most prevalent ore mineral in Çorak is pyrite it is present in all the samples. It is followed by chalcopyrite, sphalerite, and galena. Compare to sulfides, sulfosalts namely tennantite-tetrahedrite are present in lesser amounts. Covellite and bornite are rare. Gangue minerals are mainly composed of quartz and calcite. Two main mineral's assemblages are prevailing whether it is from stockwork (disseminate and veinlet zone) or from the high grade sulfide zone. In the stockwork zone pyrite has euhedral to subhedral shape. Chalcopyrite is anhedral and replaces very often pyrite. In the massive ore zone, pyrite and chalcopyrite are accompanied by large sphalerite and galena grain. Galena always crosscut sphalerite, suggesting that it is later than sphalerite. Tennantite and tetrahedrite assemblage replace very often pyrite and chalcopyrite in this zone. The pyrargyrite is present in very small amount. The tennantite is dominant in Çorak's polymetallic sulfide zone. It has been also observed galena replacing gangue minerals quartz suggesting symplectic intergrowths of ore minerals with silicate. Colloform texture is observed but they are rare. Except the pyrite, all the minerals are anhedral. Minerals size span from 50 to 300 micron. The sulfosalts are the smallest ones (Figs. 1a–f).

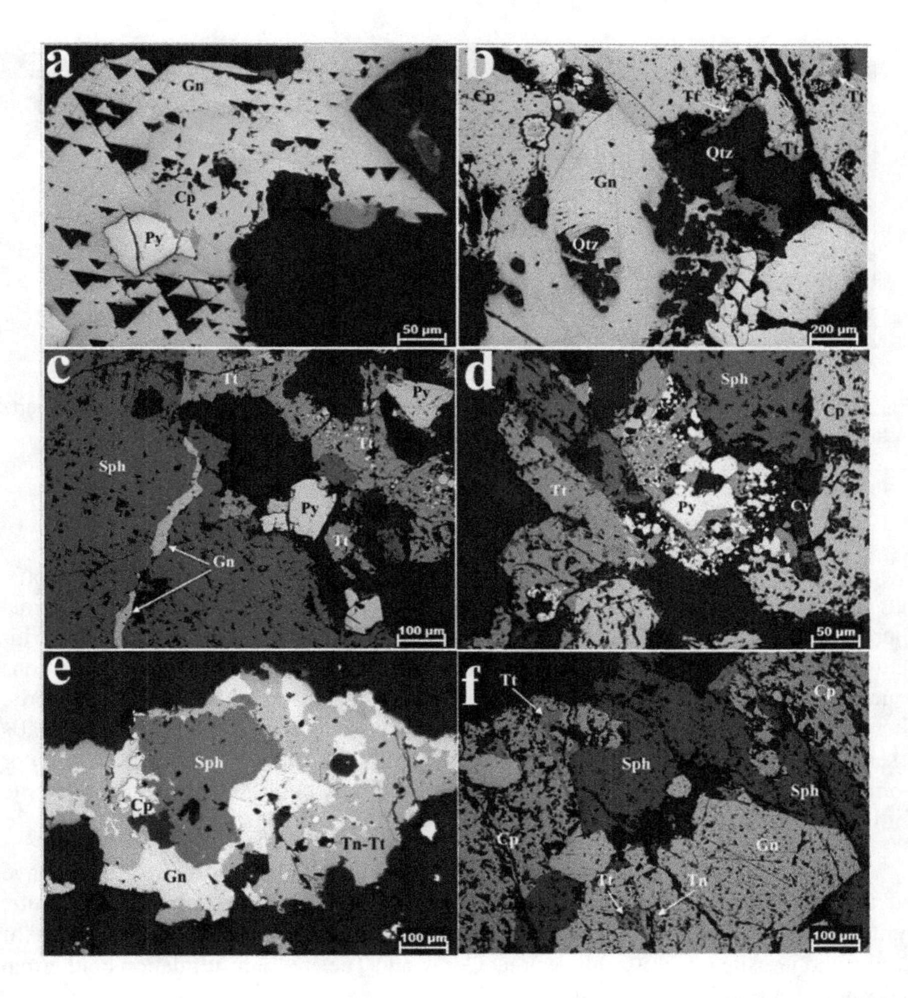

Fig. 1. (a) galena replacing chalcopyrite; (b) galena replacing quartz, replacement of gang minerals by ore minerals (symplectic intergrowths) and tennantite replacing chalcopyrite; (c) galena veins sphalerite, while pyrite replaced by tennantite; (d) tennantite replacing pyrite and chalcopyrite; (e) tetrahedrite veining galena and tetrahedrite-tennantite replacing sphalerite (Py: galena; Cp: chalcopyrite; Sph: sphalerite; Tn-Tt: Tennantite-Tetrahedrite; Qtz quartz)

The following paragenesis is deducted from the ore microscopy study: **Pyrite – Chalcopyrite (I) – Sphalerite – Galena – Sulfosalts- Chalcopyrite (II) - Bornite – Covellite**.

The cathodoluminescence study on sphalerite has revealed heterogeneity in the composition of the sphalerite (Fig. 2). These states that those sphalerite grains are originated from more than one generation phase and/or from fluids with changing composition.

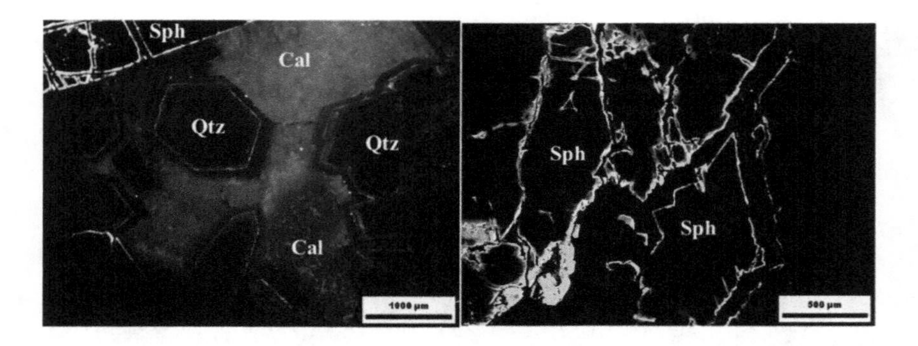

Fig. 2. Optical cathodoluminescence microscopy images (OCLMI) of sphalerite and quartz grains with infilling late calcite, showing compositionally heterogeneity sphalerites

4 Conclusions

Taç and Çorak deposits occur in north easternmost tip of the eastern Pontides tectonic belt of Turkey. They are two separate ore bodies within the same system. Hydrothermal alteration is intense and extensive in the form of silicification, epidotization, chloritization, seritization, and finally a calcification as the final stage. Three main alterations zone are succeeding from the proximal, intermediate to distal zone: Quartz + Chlorite alteration, quartz + epidote alteration, quartz + sericite alteration. Two main mineral's assemblages are prevailing whether it is from stockwork (disseminated and veinlet) or from the high grade sulfide zone. They are respectively: pervasive pyrite, chalcopyrite with sphalerite and galena.

Argillic, propylitic and silicic alteration induced by acidic hydrothermal fluids have led to the formation of quartz, muscovite, orthoclase, gypsum, dolomite, kaolinite, pyrite, sphalerite, galena and the pair jarosite-alunite assemblage. The presence of alunite and jarosite minerals indicate that Çorak and Taç are high sulfidation epithermal deposit with mesothermal signatures.

Acknowledgements. This study was financially supported by BAP Project unit (No: 2017/80) of Kocaeli University in Turkey. We are thankful to the university and to Santral Mining Co. and their employees in Yusufeli for their courteous support for the field work.

References

Bogdanov B (1980) Massive sulphide and porphyry copper deposits in the Panagjurishte district, Bulgaria. In: Jankovic S, Sillitoe Richard H (eds) European copper deposits; Proceedings of an international symposium. Springer, Berlin-Heidelberg-New York, pp 50–58

Ciftci E (2000) Mineralogy, paragenetic sequence, geochemistry, and genesis of the gold and silver bearing upper cretaceous mineral deposits, Northeastern Turkey. PhD dissertation, University of Missouri-Rolla, Missouri, USA, p 251

Kouzmanov K, Moritz R (2009) Late Cretaceous Porphyry Cu and epithermal Cu-Au association in the Southern Panagyurishte District, Bulgaria: The paired Vlaykov Vruh and Elshitsa deposits

Muntean JL, Einaudi MT (2000) Porphyry gold deposits of the Refugio district, Maricunga belt, Northern Chile. Econ Geol 95:1445–1472

19

New Data on Microhardness of Placer Gold

Z. Nikiforova(✉)

Diamond and Precious Metal Geology Institute SB RAS, Yakutsk, Russia
znikiforova@yandex.ru

Abstract. Microhardness of eolian gold – new morphologic type of placer gold, is studied. Flaky gold particles with a elevation at the periphery, as well as gold of toroidal and globular-hollow forms belong to eolian gold. Genesis of eolian gold is related to mechanical transformation of flaky gold particles into toroidal, and then into globular-hollow form in eolian conditions, that is experimentally proven. Previous studies determined changes of microhardness, mainly from 47 kg/mm^2 to 100 kg/mm^2, lower limit – 40 kg/mm^2. But, low microhardness was identified in globular-hollow gold for the first time, which stood at 21 kg/mm^2. This is due to the fact that, as a result of transformation of flaky gold in eolian conditions, under mechanical and chemical processes, silver and trace elements were removed, that led to fineness increase up to 1000‰, and to decompaction of inner structure of gold, that influenced microhardness indices. Identified patterns in nature, microhardness changes under mechanogenic and chemical process impact in eolian conditions, can be successfully used in gold metallurgy to produce gold alloys with very low microhardness.

Keywords: Gold · Microhardness · Fineness · Trace elements · Morphology · Process

1 Introduction

Microhardness of gold was earlier studied by Lebedeva (1963), Badalova et al. (1968), Petrovskaya (1968), Popenko (1982). Previous researchers identified that, microhardness changes on the average from 40 to 100 kg/mm^2 and depends on chemical composition of gold (percentage of silver and trace elements) and its inner structure. It is known that, Ag and Cu trace elements significantly increase hardness of gold alloys. Trace element presence (Pt, Sn, Al) also causes sharp increase of gold hardness. Gold of low fineness (550–650‰) has the biggest microhardness (Popenko 1982).

2 Methods and Approaches

Microhardness of globular-hollow eolian gold was studied by microhardness tester PMT-3. In order to detect microhardness, globular-hollow forms of gold were mounted in epoxy specimen. In such a manner, optimum section of gold particles was obtained, necessary for further study; it is possible to identify microhardness indices in its central part, end and shell. 55 identifications were performed at 12 sections of globular-hollow gold particles (Fig. 1 and Table 1). Since, gold fineness is increased in eolian conditions, high-standard shell is formed in end parts, and central parts remain unchanged; microhardness was measured in end parts of the globular-hollow gold, and in partition of hollow ball, being a relic of the flake.

3 Results and Discussion

Microhardness of native gold, being in eolian conditions, is studied for the first time. It is identified that, in four gold particles (1, 2, 3, 4), increase of microhardness from partition center to the end is clearly observed. Microhardness between individual grains of gold particles (1, 2) is 2 times different. Uniform microhardness is typical for different parts of the globular-hollow forms (6, 7, 9). Sample 12 yields increase of microhardness from outer part of the end to partition center. Other four gold particles do not have any clear regularity concerning microhardness distribution.

When microhardness of eolian gold was studied, it was found that, microhardness depends on degree of inner structure changes and chemical composition of gold in exogenetic conditions. When flaky gold is transformed in a complicated way into toroidal form, and then into globular-hollow form in eolian conditions, not only gold morphology is changed, but also chemical composition is changed, in particular gold fineness is increased and amount of trace elements is decreased (Nikiforova 1999). It is explained by the fact that, as a result of complicated transformation of flaky gold in eolian conditions, very thin gold films are stretched from the flake end and overlap each other, generating globular-hollow form. In this connection, surface for chemical impact is increased in exogenetic conditions, that contributed to gradual removal of silver and trace elements, and sharp decrease of microhardness indices of eolian gold.

In addition, earlier unknown lower limit of microhardness of the globular-hollow gold is identified - 21 kg/mm^2.

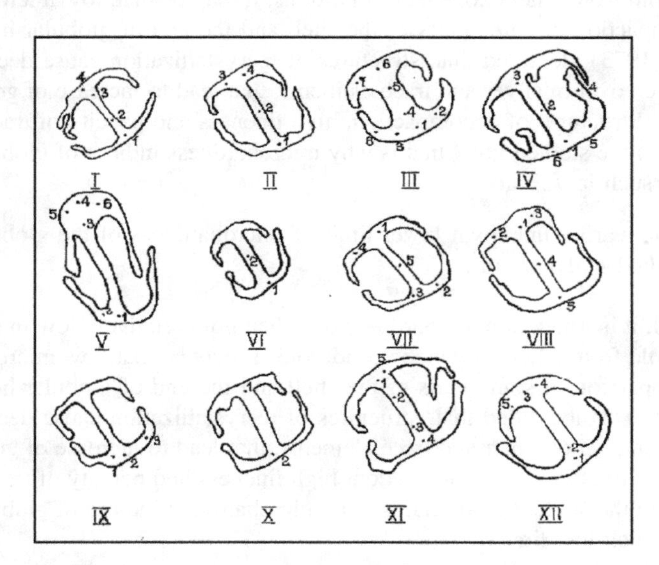

Fig. 1. Location of the points of identification of globular-hollow gold microhardness

Table 1. Microhardness of the elements of globular gold section, kg/mm^2

Sample №	Elements of section of globular form					
	End 1	Partition	End 2	Base of the		Shell
				End 1	End 2	
I	$\frac{64,5^*}{4}$	$\frac{41,2}{2,3}$	$\frac{55}{1}$			
II	$\frac{25,7}{3,4}$	$\frac{21}{2}$	$\frac{28,6}{1}$			
III	$\frac{41,2}{6}$	$\frac{25,7}{5,4}$	$\frac{32,1}{1}$	$\frac{55}{7}$	$\frac{41,2}{2}$ $\frac{25,7}{3}$	$\frac{23,2}{8}$
IV	$\frac{46,2}{1}$	$\frac{42,0}{2,3}$		$\frac{64,2}{6}$		$\frac{57,2}{5}$ $\frac{46,7}{4}$
V	$\frac{47,3}{4}$	$\frac{32,1}{3}$ $\frac{25,7}{2}$	$\frac{25,7}{1}$	$\frac{55}{5}$ $\frac{23,2}{6}$		
VI	$\frac{43,0}{1}$	$\frac{41,2}{2}$				
VII	$\frac{64,5}{4}$	$\frac{64,5}{5}$	$\frac{64,5}{2}$	$\frac{64,5}{3}$	$\frac{41,2}{1}$	
VIII	$\frac{47,3}{1}$	$\frac{47,3}{4}$	$\frac{32,1}{5}$	$\frac{32,1}{2,3}$		
IX	$\frac{28,6}{1}$			28,6		$\frac{28,6}{3}$
X	$\frac{47,3}{1}$	$\frac{55,0}{2}$				
XI	$\frac{36,2}{1}$	$\frac{47,3}{2}$ $\frac{36,2}{3}$	$\frac{36,2}{4}$	$\frac{41,2}{5}$		
XII	$\frac{47,3}{3}$	$\frac{55}{2}$	$\frac{41,2}{1}$	$\frac{36,2}{4}$	$\frac{32,1}{5}$	

*Above line – microhardness, below line – point of microhardness measurement (Fig. 1).

In general, it is true that microhardness of eolian gold relatively low in comparison with native gold from other exogenetic conditions. It may be that, low microhardness is due to decompaction of some parts of the shell and the end of globular-hollow gold. Petrovskaya (1973) identified that, structures of recrystallization cause decompaction, stress relief, removal of silver and trace elements, that lead to increase of gold fineness up to 1000‰. Structures of decompaction, high fineness and paucity of trace elements are observed in the studied gold, that is why microhardness indices of globular-hollow gold reached such low limits.

In addition, earlier unknown lower limit of microhardness of the globular-hollow gold is identified - 21 kg/mm^2.

In general, it is true that microhardness of eolian gold relatively low in comparison with native gold from other exogenetic conditions. It may be that, low microhardness is due to decompaction of some parts of the shell and the end of globular-hollow gold. Petrovskaya (1973) identified that, structures of recrystallization cause decompaction, stress relief, removal of silver and trace elements, that lead to increase of gold fineness up to 1000‰. Structures of decompaction, high fineness and paucity of trace elements are observed in the studied gold, that is why microhardness indices of globular-hollow gold reached such low limits.

4 Conclusions

Thus, low microhardness – 21 kg/mm^2 is identified in globular-hollow gold is identified for the first time. Identified patterns in nature, microhardness changes under mechanogenic and chemical process impact in eolian conditions, can be successfully used in gold metallurgy to produce gold alloys with very low microhardness.

Acknowledgements. The work is implemented within scientific-research projects of Diamond and Precious Metal Geology Institute, Russian Academy of Sciences, project № 0381-2016-0004.

References

Badalova RP, Nikolaeva EP, Tolkacheva LF (1968) Study of microhardness of minerals of the silver-gold series from gold deposits of Uzbekistan. "Physical features of rare-metal minerals and methods of their study" collected articles, Moscow, Nauka

Filippov VE, Nikiforova ZS (1988) Transformation of native gold particles during process of eolian impact. AN SSSR Report, vol 299, № 5, pp 1229–1232

Lebedeva SI (1963) Identification of microhardness of minerals. AN SSSR Publishing House, Moscow

Nikiforova ZS (1999) Typomorphic features of eolian gold. ZVMO, N5, pp 79–83

Petrovskaya NV (1973) Native gold. Nedra, Moscow

Popenko GS (1982) Mineralogy of fold from the Quaternary deposits of Uzbekistan, Tashkent, FAN

20

Modal Analysis of Rocks and Ores in Thin Sections

Yu. Voytekhovsky(✉)

Saint-Petersburg Mining University, Saint-Petersburg, Russia

Voytekhovskiy_YuL@pers.spmi.ru

Abstract. The article is devoted to the history and justification of the modal analysis of rocks and ores with a microscope. It is shown that the Delesse-Rosiwal-Glagolev ratios do not follow from the Cavalieri principle. They do not allow one to find the exact volume of the minerals in rocks or ores, but give only their average estimates. It is also shown that the volume fractions of convex mineral grains in rocks and ores, taken equal to the fractions of their flat sections, are always underestimated if compared with the matrix. Due to the wide variety and complexity of forms of mineral grains, the methods of stereological reconstruction lead to integral equations with a difficult to define form factor. Most likely, tomography methods should come to replace the modal analysis of rocks and ores in thin sections.

Keywords: Rocks and ores · Modal analysis · Stereological reconstruction

1 Introduction

Modal (quantitative mineralogical) analysis of rocks and ores in thin sections is one of the first fundamental quantitative methods of mineralogy (including technological mineralogy), petrography (in classifications of rocks and ores, and petrological reconstructions), and lithology (i.e., petrography of sedimentary rocks). That is why its rigorous justification is of fundamental importance owing to the Delesse-Rosiwal-Glagolev ratios, as well as to stereological reconstruction. It makes sense to summarize the history of these methods in Russia and abroad, and to formulate the conclusion about their prospects.

2 The Cavalieri Principle and the Delesse-Rosiwal-Glagolev Ratios

First of all, we point out that the relations suggested by Delesse (1848) $dV_i = dS_i$, Rosiwal (1898) $dS_i = dL_i$, and Glagolev (1932) $dL_i = dN_i$, decreasing the dimension of space (namely, equating the volume fractions of minerals to areal, areal to linear, linear to point-like), have no relation to the Cavalieri principle: $S_{1i} = S_{2i} \rightarrow V_1 = V_2$ (if the areas of all arbitrarily close parallel sections of two bodies are pairwise equal, then their volumes are also equal, Fig. 1).

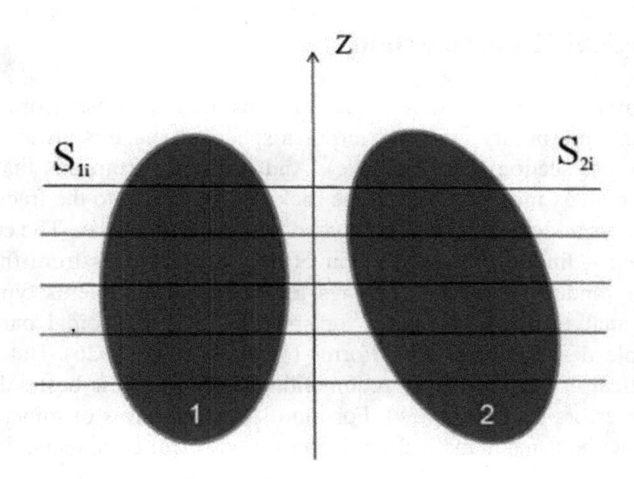

Fig. 1. To the justification of the Cavalieri principle

In recent notation, this principle, historically preceding integral calculus, has a clear meaning: $V_1 = \int S(z)\,dz = V_2$, where S(z) is a continuous function of the areal fraction of a certain mineral along the z axis being normal to the sections. But, the modal analysis of rocks and ores in thin sections, which accumulates the statistics of areal, linear or point fractions of minerals from section to section, has nothing to do with the integration procedure. It only leads to an assessment of their average values. In this case, it can be argued that the volume of any mineral in a rock or ore is within the certain interval:

$$S(z)_{min}\Delta z = \int S(z)_{min} dz < V < \int S(z)_{max} dz = S(z)_{max}\Delta z$$

where Δz is the thickness of rock sample under study.

Despite of this contradiction, which was also considered in the works (Krumbein 1935; Chayes 1956), the method became firmly established in practice because of its apparent simplicity and was step by step automated (Shand 1916; Wentworth 1923; Hunt 1924; Dollar 1937; Hurlbut 1939) up to the use of modern computers for image analysis of thin sections. The list of parameters characterizing the cross-section of minerals, and the speed of processing have grown many times. But, in terms of the reconstruction of the true metric characteristics of mineral grains from those of their flat or even linear sections, the ideology remains the same.

Companies that produce computer structure analyzers offer software packages without discussing the fundamental problems. The analysis of 2D images does not use the available chapters of mathematics. For example, the distances between the mineral grains in thin section are replaced by the Euclidean distances between the points taken within the grains, whereas there is a more complicated, but easily programmable Hausdorff's metric which allows one to do this procedure correctly. In turn, it gives us the possibility of calculating space covariograms between the mineral grains of different species and their various clusters in the rock.

3 Stereological Reconstruction

A new line of research, i.e. stereological reconstruction, arose from the obvious observation that an arbitrary cross section of a spherical shape is always less than its characteristic cross section (Fig. 2, above). And it follows from this that the volume fraction of the convex mineral phase in the rock and ore, equal to the fraction of its flat sections, is always underestimated if compared with the host matrix. The corresponding general problem – finding the distribution of true particle sizes from the size distribution of their random sections – belongs to the inverse problems typical for geophysics. It is analytically solved only for spherical and ellipsoidal particles due to relatively simple description of these forms (Wicksel 1925, 1926). But, even in this case, the practical use of the theory requires the selection of the best solution and an estimate of the errors (Fig. 2, below). For more complex forms of mineral grains this can't be done without mathematical modeling on powerful computers.

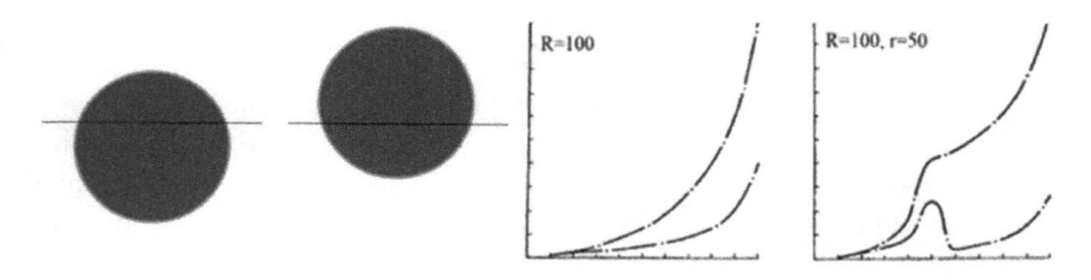

Fig. 2. Left: the size of the cross section of a convex grain is always less than the characteristic one, in the modal analysis its volume fraction is underestimated. Right: to recognize the true size of spherical particles by the size of circular sections; R = 100 – particles with a radius of 100 arbitrary units; R = 100, r = 50 – two sets of particles of the same type, indistinguishable in sections (for example, two generations of one mineral); horizontal scale – section radii from 0 to 100 bits per 10 classes; vertical scale – frequencies by classes (lower curves) and accumulated frequencies (upper curves).

The history of this area in Russia can be found in the following incomplete list of works: Zhuravsky A.M. Mineralogical analysis of thin section in terms of probabilities. Moscow-Leningrad: Gosgeolizdat, 1932. 20 p.; Glagolev A.A. Quantitative mineralogical analysis of rocks with the microscope. Leningrad: Gosgeolizdat, 1932. 25 p.; Glagolev A.A. On the geometric methods of quantitative mineralogical analysis of rocks. Moscow-Leningrad: Gosgeolizdat, 1933. 47 p.; Glagolev A.A. Geometric methods for quantitative analysis of aggregates with a microscope. Moscow-Leningrad: Gosgeolizdat, 1941. 263 p.; Chayes F. An elementary statistical appraisal. New York: John Wiley & Sons, Inc., 1956; Shvanov V.N., Markov A.B. Granulometric analysis of sandstones in thin sections//Geology and exploration. 1960. N 12. P. 49–55; Ivanov N. V. A new direction in testing ore deposits. Moscow: Gosgeolizdat, 1963. 179 p.; Chernyavsky K.S. Stereology in metallurgy. Moscow: Metallurgy, 1977. 375 p.; Ivanov O.P., Ermakov S.F., Kuznetsova V.N. Improving the accuracy of determining the weight particle size distribution of ore minerals from measurements in thin sections//Proc. CNII of Tin. Novosibirsk: Science, 1979. p. 10–14; Gulbin Yu.L. On stereological reconstructions of grain sizes in aggregates//Proc. Rus. Miner. Soc. 2004. N 4. P. 71–91.

4 Conclusions

Thus, due to the extraordinary diversity and complexity of the forms of mineral grains in rocks and ores, the methods of stereological reconstruction lead to integral equations with an analytically difficult-to-define form factor. The practical application of the theory is drowning in the selection of the best solution to the inverse problem and complex estimates of measurement errors. It seems that the modal analysis of rocks and ores in thin sections should be replaced by tomography methods. Standardizing modal analysis of rocks and ores in thin sections by creating their artificial counterparts with previously known volume fractions of mineral grains and a wide range of petrographic structures can serve as an inter-laboratory comparison of the accuracy of the method. But it does not solve the problems in essence.

References

Delesse M (1848) Procede mecanique pour determiner la composition des roches. In: annales des mines. De memoires sur l'exploitation des mines. 4me serie. T. XIII. Carilian-Goeury et Dalmont, Paris, pp 379–388

Dollar ATJ (1937) An integrating micrometer for the geometrical analysis of rocks. Mineral Mag 24:577–594

Hunt WF (1924) An improved Wentworth recording micrometer. Am Mineral 9:190–193

Hurlbut CS Jr (1939) An electric counter for thin-section analysis. Am J Sci 237:253–261

Krumbein WC (1935) Thin-section mechanical analysis of indurated sediments. J Geol 43:482–496

Rosiwal A (1898) Über geometrische Gesteinanalysen. Ein einfacher Weg zur ziffermässigen Feststellung des Quantitätsverhältnisses der Mineralbestandtheile gemengter Gesteine. In: Verh. der k.-k. Geol. Reichsanstalt. Verlag der k.-k. Geol. Reichsanstalt, Wien, pp 143–175

Shand SJ (1916) A recording micrometer for geometrical rock analysis. J Geol 24:394–404

Wentworth CK (1923) An improved recording micrometer for rock analysis. J Geol 31:228–232

Wicksel SD (1925) The corpuscle problem: a mathematical study of a biometric problem. Biometrica 17:84–99

Wicksel SD (1926) The corpuscle problem: case of ellipsoidal corpuscles. Biometrica 18:151–172

21

Development of Methods for Anti-Filtration Formations Destruction Inside a Heap Leach Pile

H. Tcharo(✉), M. Koulibaly, and F. K. N. Tchibozo
Department of Mineral Developing and Oil&Gas Engineering, Engineering Academy, RUDN University, Moscow, Russia
honoretcharo@yahoo.com

Abstract. This article discusses the new technical solutions that increase the restoration of the quality of pregnant solutions flowing out from the heap leach pile.

Keywords: Anti-filtration formations · Heap pile · Pipe · Sprinklers

1 Introduction

The presence of low permeable soils is one of the most difficult problems of metal extraction during heap leaching (Ozhogina et al. 2017; Vorobyov et al. 2017).

The development of new models will reduce the degree of their negative impact on the intensity of heap leach process.

2 Methods and Approaches

A wide range of methods were used for the current research: the analysis of the earlier conducted researches, the mathematical and physical simulation, determination of dependences, the calculations and the control, the experimental studies and measurements in accordance with the conventional standard.

For the experience, we used a section of a rectangular shaped heap leach pile 5 × 5 m size. We have installed an air injection pipe near a homogeneous low-permeable layer. Assuming that the bottom and sides are sealed to air leakage, we used the SVAirFlow software to simulate the air/oxygen flow through a uniform anti-filtration layer. In addition, the temperature inside the mass is constant and equal to 20 °C. The air supply pressure through the pipeline is 121 kPa, and the upper part has an atmospheric pressure of 101 kPa.

3 Results and Discussion

The results of the model are shown in Fig. 1. We've noticed that the air pressure in the whole low-permeable layer varied between 109 and 120 kPa. This means that the pores are slightly wider open.

The second technical solution, which also allows to increase the intensification of heap leach process is the displacement of spraying devices in the direction of weakly affected by technological solutions places.

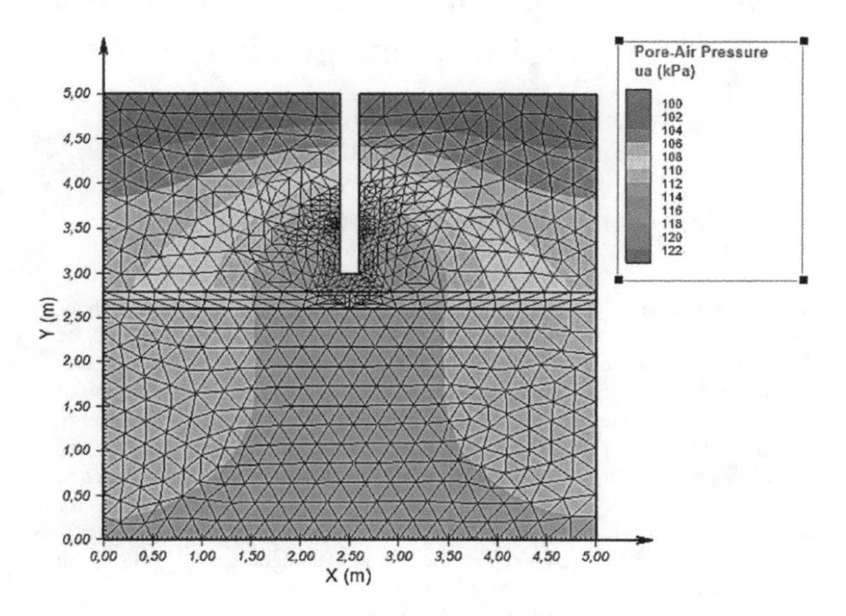

Fig. 1. Model showing the impact of injecting air/oxygen

4 Conclusions

According to our study results, we can conclude that supplying the air under much higher pressure than the atmospheric one in the places where anti-filtration layers are formed, along with the displacement of the sprinklers, will enhance more seepage of pregnant solutions through a heap leach pile.

References

Ozhogina EG, Shadrunova IV, Chekushina TV (2017) Mineralogical rationale for solving environmental problems of mining regions. Gornyi zhurnal 11:105–110

Vorobyov AE, Lyashenko VI, Tcharo H, Vorobyev KA (2017) Efficiency increase for gold-bearing ore deposits with respect to the influence of sulfide surface nanobarrier on metal adsorption. Sci Tech J "Metall Min Ind" 7:28–36

22

Predictive Assessment of Quality of Mineral Aggregates Disintegration

S. Shevchenko, R. Brodskaya(✉), I. Bilskaya, Yu. Kobzeva, and V. Lyahnitskaya

Karpinsky Russian Geological Research Institute, St. Petersburg, Russia
Rimma_Brodskaya@vsegei.ru

Abstract. The paper is supposed to discuss one of the most important technological properties of mineral aggregates: their ability to disintegrate, to be destroyed for the subsequent redistribution and extraction of a useful component. The strength of a mineral aggregate is determined by the strength of accretion of composing individuals and depends on many factors, including the energy (kinetics) of formation conditions and the subsequent transformation conditions. The energy of the boundary of accretion of mineral grains in aggregate cannot be directly measured since currently, there are no reliable and proved methods for measuring the surface energy of solids. It is only possible to calculate quantity, proportional to the surface energy of the fusion of mineral grains – this is the calculation of the atomic density of the surface, parallel to the boundary of accretion of a pair of mineral grains. The methodology and calculation of this characteristic have been developed and published. The use and obtaining of results are possible at the stage of preliminary mineralogical study by methods of geometrical analysis of the structure of the mineral aggregate. The effectiveness of the proposed method is determined by the knowledge of the quality and features of destruction of the aggregates using different physical methods of disintegration: traditional mechanical, ultrasonic, electro-pulse, electro-hydraulic.

Keywords: Technological properties · Mineral aggregate · Aggregate disintegration · Repartition · Strength · Accretion boundaries · Energy of boundaries

1 Introduction

The mineral aggregate is considered as a structured system consisting of mineral individuals and the accretion boundaries between them. The formation of the aggregate results from cooperative thermodynamic processes of synthesis of substances, corresponding to the composition of minerals, as well as the crystallization of mineral individuals and the formation of crystal-grain boundaries, their aggregation. The aggregation or accretion of mineral individuals with each other is carried out along the forming boundaries of accretion. The process of aggregation of grains in the mineral aggregate may somewhat lag behind the synthesis and crystallization of individuals, but their parallel development is also possible. The accretion boundaries do not always correspond to the boundaries of the crystallization of individuals, since they are formed in a different energy situation.

2 Methods and Approaches

The accretion boundaries of mineral individuals represent a certain transition zone between crystal lattices of accreted grains. The orientation of the accretion boundaries with respect to the symmetry of crystalline lattices of mineral grains depends on the local energy potentials during the formation and transformation of the mineral individuals and the aggregate. Local thermodynamic potentials of the forming mineral system directly affect the safety of an individual through the energy of its boundary, the choice of the orientation of the boundary relative to the symmetry of its crystal lattice and the orientation of the accretion boundary with the nearest stable grain. Thus, a compromise boundary is established for the accretion of each pair of mineral individuals, which is stable in a specific range of thermodynamic forces and flows.

During evaluation of sequence and quality of opening of the accretions, it is necessary to take into account the distribution of mineral individuals in the aggregate which possess a very perfect cleavage.

It is possible to assess the degree and quality of opening of the accretions at the stage of preliminary mineralogical analysis when studying the structure of the mineral aggregate in thin sections using geometric analysis methods.

3 Results and Discussion

The boundaries of mineral individuals, as well as the accretion boundaries, possess a certain stability reserve, ensuring their functioning in a certain range of thermodynamic and mechanical parameters. The stability margin of the boundaries is the wider, the higher the kinetics of the cooperative processes that form them, their genesis. The accretion boundaries of mineral grains are the more stable in the field of destructive forces, the more surface energy they accumulate.

The stability of the accretion boundary of mineral grains depends, in addition to the above said, on the degree of its balance by the value of energy saturation of each of the mineral grains in the accretion region.

The open boundaries of the mineral grains of accretions after relaxation of the crystal lattice, as well as remaining unopened accretions, have different elemental composition, amount of energy, and different floatability.

4 Conclusions

The accretion boundaries determine the grain system as a mineral aggregate and determine its stability, as both the strength of the accretions of mineral grains, and the strength of the mineral aggregate as a whole.

The strength of mineral aggregates is understood as resistance to destructive forces on the system of mineral grains forming the aggregate due to the presence of their "accretion" surfaces.

The accretion boundaries of the mineral grains can be balanced or unbalanced by value of energy of accretion surfaces of individuals.

The accretion of grains on surfaces, which atomic density or crystallographic indices correspond to the edges of habit forms, will collapse faster and earlier by the same destructive force, than the accretion of energy-intensive boundaries with a low atomic density and high crystallographic indices.

References

Brodskaya RL, Makagonov EP (1990) Determination of the spatial orientation of crystalline individuals in microstructural analysis and prospects for its use in stereometric petrography. Notes All-Union Mineral Soc 119(4):84–93

Brodskaya RL, Bil'skaya IV, Lyakhnitskaya VD, Markovsky BA, Sidorov EG (2007) Boundaries of accretions between mineral individuals in an aggregate. Geol Ore Deposits 49(8):669–680

23

Phenomenon of Microphase Heterogenization by Means of Endocrypt-Scattered Impurity of Rare and Noble Metals as a Result of Radiation by Accelerated Electrons of Bauxites

I. Razmyslov[1(✉)], O. Kotova[1], V. Silaev[1], and L. A. Gomze[2]

[1] Institute of Geology Komi SC UB RAS, Syktyvkar, Russia
razmyslov-i@mail.ru
[2] University of Miskolc, Miskolc, Hungary

Abstract. During radiation-thermal transformation of Timan ferrous bauxites we discovered a previously unknown phenomenon of microphaseheterogenization, which can contribute to the extraction of many valuable impurities during processing of relatively poor quality bauxite raw.

Keywords: Bauxite · Radiation-thermal transformation · Microphase heterogenization · Profitability of bauxite raw processing

1 Introduction

The problem of processing of aluminum ores is related to the fact that bauxite-forming minerals are characterized by similar values of density, dispersion and fine mutual intergrowths of individuals, non-recoverability of many valuable microelements. Therefore, the development and improvement of methods for the enrichment and processing of bauxite remains highly relevant (Borra et al. 2015; Borra et al. 2016; Davros et al. 2016). In terms of their iron index, the studied Timan bauxites are subdivided into three mineral types: hematite-boehmite, hematite-berthierine-boehmite, and berthierine-boehmite (Vakhrushev 2011, 2012; Vakhrushev et al. 2012). The results of our studies showed that in these bauxites in the state of endocrypt scattering there are a lot of valuable elements-impurities, which extraction by modern technologies is either time consuming or not yet possible. Our experiments on heating of bauxites in combination with their irradiation with high-energy electrons lead to a change in the phase composition of bauxites and, as a result, to the improvement of their technological properties (Rostovtsev 2010; Kotova et al. 2016).

2 Methods and Approaches

We carried out experiments with thermal (heating to 500 and 600 °C with quadruple exposure by 60 min) and radiation-thermal (heating to 500–600 °C with double exposure by 20 min with irradiation by an electron beam with energy of 2.4 MeV using ILU-6 industrial unit at the Institute of Nuclear Physics of the Siberian Branch of the Russian Academy of Sciences) modification of Timan iron bauxites.

3 Results and Discussion

Heating up to 500–600 °C with exposure to 60 min without irradiation led to almost complete dehydration of Al and Fe oxyhydroxides with the formation of γAl_2O_3 (spinelide with a defective structure) and hematite with relatively small alteration of structure and magnetic properties of the rocks. The gross chemical composition of the annealed samples remained almost unaltered, and the content of trace elements increased markedly in% to the original: Cu by 70–250; Zn by 20–25; Cd by 20–380; Zr by 2–20; Nb by 15–20; Sc by 25–40; Y by 35–70; Ce by 25–60; Nd by 1–10. Obviously, the latter is conditioned bya significant thermal dehydration of the studied bauxites.

The radiation-thermal treatment of ferrous bauxites led not only to dehydration of the original minerals, but also to chemical disproportionation of the original substance, its radical magnetic restructuring resulted from the presence of strong magnetic phases (maghemite, magnetite) and the formation of valuable trace elements of many new minerals with individuals varying in shape from isometric to needle-like and in size from submicronic to 0.5 mm due to endocrypt scattered impurity. The newly formed phases include native metals (Au, Pb, Al, Cu, Zn); sulfides (pyrite, galena); oxides of Sn, Ta, Nb, Zr, lanthanides; silicates (zircon, kaolin); rare sulphates, etc. Thus, the heating, combined with irradiation with high-energy electrons, resulted not only in transformation of primary minerals in the ferrous bauxites, but also in microphase heterogenization with the formation of new minerals (Fig. 1). It can be assumed that this kind of transformation can contribute to extraction of many valuable impurities and increase the profitability of processing of relatively low-quality bauxite raw.

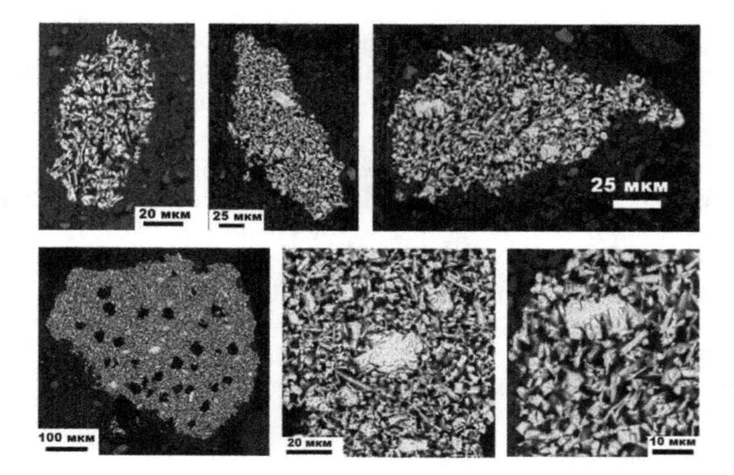

Fig. 1. Micro pocket segregations of Ce-Gd oxides in radiation-thermally modified Timan bauxites

Acknowledgements. This research was supported by UB RAS project 15-18-5-44 and project AAAA-A19-119031390057-5 "The main directions of integrated assessment and effective use of geo-resources in the Timan-North Ural-Barents Sea region".

References

Borra CR, Mermans J, Blanpain B, Pountikes YB, Gerven T (2016) Selective recovery of rare earths from bauxite residue by cjmbination of sulfation, roasting and leaching. Min Eng 92:151–159

Borra CR, Pontikes Y, Binnemans K, Gerven T (2015) Leaching of rare earts from bauxite residue (red mud). Min Eng 76:20–27

Davros P, Balomenos E, Panias D, Paspaliaris I (2016) Selective leaching 0f rare earth elements from bauxite residue (red mud). Hydrometallurgy 164:125–135

Kotova OB, Razmyslov IN, Rostovtsev VI, Silaev VI (2016) Radiation-thermal modification of ferruginous bauxites during processing. Enrich Process 4:16–22 (in Russian)

Rostovtsev VI (2010) Theoretical foundations and practice of using electrochemical and radiation (accelerated electrons) effects in the processes of ore preparation and enrichment of mineral raw materials. Vestnik of the Chita State University, vol 8, pp 91–99. (in Russian)

Vakhrushev AV (2012) Crystal chemistry of bauxite minerals from the Vezhayu-Vorykvinskoe deposit. Structure, substance, history of the lithosphere of the Timan-Northern Ural segment. Geoprint, Syktyvkar, pp 32–34. (in Russian)

Vakhrushev AV, Lyutoev VP, Silaev VI (2012) Crystal-chemical features of ferrous minerals in bauxite at the Vezhayu-Vorykvinskoe deposit (Middle Timan). IG Komi Science Center UB RAS, pp 14–18. (in Russian)

24

Mineral Preparation in Geological Research

T. Yusupov[(✉)], A. Travin, S. Novikova, and D. Yudin
V.S. Sobolev Institute of Geology and Mineralogy SB RAS, Novosibirsk, Russia
yusupov@igm.nsc.ru

Abstract. Paper deals with discussion of minerals' preparation requiring high purity monoproducts, this has special importance for minerals –geochronometers. Widely used methods including gravitation, magnetic separation, floatation provide fractions with 90% of targeted mineral. Further monomineralitiy increase requires special separation methods; one of them – "Strat" is perspective. It is based on separation in organic liquids under gradual density change. Combination of bromoform with d – 2.89 g/sm^3 and dimethyle formamide with d – 0.8 g/sm^3 is used; density gradations till 0.001 g/sm^3 are possible therefore isomorphic inclusions could be separated. Another direction is presented by trybotreatments under higher energies in planetary mills with a centrifugal factor to 40–50 g. Exotic surface substances presented mainly by kaolinite, muscovite, calcite, gothite are removed as trybotreatment result. Special planetary mills - classifiers are used for processing of big samples. This method together with minerals opening in disintegrator under destruction by the free pulse is recommended for wide application.

Keywords: Monomineralitiy · Geo chronometers · Separation · Surface · Organic liquids · Trybotreatment

1 Introduction

Many mineralogical and geochemical researches are based on mono mineral products' studies; acquisition of these products is based on research intensive processes of minerals' revealing and extracting. This assumption is relevant to geochemical, geophysical, lithologic, petrochemical and other studies; geochronological definitions became wider during recent years.

The task to extract minerals – chronometers of mono mineral purity is very important and complicated. The problem becomes much more complicated when chronometers are extracted from geo objects which have precious and rare metal character. This happens due to their very low content, thin dispersion and occurrence in genetic association (intergrowth) with usually rock-forming minerals.

Range of rock-forming and ore minerals which are used for rock dating is widening. Analysis is concentrated on such minerals as plagioclases, olivine, ortho and clino pyroxene, phlogopite, tourmaline, sphalerite, volframite, tin spar, pyrite, pyrrhotite, pentlandite and others. Special methods should be developed for the extraction of many of these minerals (Isotope…, 2015; Methods…, 2018).

2 Methods, Approaches, Results and Discussion

We develop new section in sample preparation – mineral preparation, which includes number of new research and methodological aspects.

1. Preliminary concentrating of minerals – chronometers from objects with very low their content. Only intermediate products which are used for mono mineral fractions extraction could be obtained by traditional types of separation – gravity, magnet and floating.

   Gravitation methods are helpfully used in the situations where differences between density of extracted mineral and monaural basis are not lower than 3 kg per sm^3. Losses of target mineral are significant under other combinations of densities. Magnet methods are more effective when differences in magnet sensitivity of separating components are sufficient for separation. This is true for the case of quartz from −0.40 to 0.10 and biotite from + 46.7 to 86.7 109 m^3/g. Under lower differences in magnet sensitivity of minerals which are contained in samples at the level of accessory units extraction is extremely difficult.

   Great perspectives in the extraction of minerals from extremely poor subsurface rocks are related with floatation process which makes it possible to get minerals with less than 0.1% content in sample. Obtained products of preliminary concentration should either be further grinded in order to open intergrowths or be dressed with the help of special concentrating methods till mono mineral state. Combine schemes with mono mineral and similar products extractions at the initial stage are often used; then intergrowths minerals opening and repeated concentrating of targeted minerals take place (Berger 1962).

2. Opening of minerals from intergrowth stage done with the help of mechanical treatments may be accompanied by significant structural chemical changes. It is necessary to avoid high temperatures, local high pressures and if possible to use dry process. High energetic free pulse realized in desintegrators is effective method of minerals – chronometers' opening. Prospectivity of desintegrator's use for ore preparation is proved on the cases of different minerals: spodumene, apatite, sulfides and others. Higher preservation of crystal structure and lower over grinding are considered to be main advantages here (Yusupov et al. 2015). Positive aspects of mineral preparation were revealed under disintegrating of quartz – feldspar associations (Yusupov et al. 2018). Disintegrated sample preparation is recommended for wide use.
3. Obtained concentrates were dressed by methods of mono mineral fractions extraction with extraction of products with 90% of targeted mineral (Methods… 1985). Further increase of mono mineral character is reached by the help of special methods. Gradual separation in organic liquids on density – "Starts" method and trybo treatment – surface attrition under higher energies of mechanical treatment are wide used. Mixture of bromform with d – 2.89 g/sm^3 and dimethyle formamide with d – 0.8 g/sm^3 is used as separation media. Different density gradations till 0.001 g/sm^3 are possible here. Potential of the method is shown on the example of quartz – feldspar associations with – 0.3 +0.2 mm size (Table 1).

Table 1. Reparability of minerals of non electromagnetic quartz feldspar product

Fraction density g/sm^3	Output, %	Elements content, %					
		SiO_2	Fe_2O_3	Al_2O_3	K_2O	Na_2O	Li_2O
Initial		81.1	0.03	12.3	2.48	5.18	0.028
2.44–2.55	7.881	65.3	0.02	19.3	13.3	1.28	0.097
2.55–2.58	3.751						
2.58–2.61	5.109						
2.61–2.63	43.795	71.0	0.03	18.4	0.51	9.82	0.004
2.63–2.65	35.737	98.7	0.03	1.4	0.12	0.62	0.017

Potassic feldspar product with 7.88% output was extracted under 2.44–2.55 g/sm^3 density. Fraction of sodium feldspar is concentrated under higher density d – 2.61–2.63 g/sm^3 with output of 43.79%. Density of quartz fraction is more close to similar indicator for sodium spar with density interval being 2.63–2.65 g/sm^3 and output 35.73%.

These results confirm high effectiveness of density method; separation of 0.05 mm and lower size products looks possible (Yusupov et al. 2015). Method is successfully used for not only quartz and feldspar separation but also for muscovite, biotite, glauconite (Katz 1977).

Trybo attrition impacts are under investigated though they are important for mineral's homogeneity increase. They enable to remove inclusions of tramp substances presented mainly by kaolinite, muscovite, calcite, gothite as well as by remaining floating reagents.

It is important to take into account that after removal of surface layers of 0.1–10 mcm thickness surface is characterized by different structural imperfections. They could vary from practically unchanged state to totally crystal and chemically destroyed surface (Yusupov et al. 2018).

Trybo treatment looks as important way to increase mono mineral character. Hand attrition in jet and jasper stamps is widely used in institute's analytical practice. Method is applied for treatment of biotite, glauconite, amphibolites, tourmolin spinels, phosphates, sulphides and other minerals with monomimeral coefficients being about 100%.

Facilities of PMK type are effective for large samples with weight more than 1 kg trybo treatment. Material here is exposed by planetary rotating movement of ore mass. Such technological regime selectively destroys impurity substances and increases monomineral properties of extracting products. Improved version of this mechanism is being developed at CJSC "Itomak", Novosibirsk.

Trybotreatment not only removes impurity substances but also changes surface defectiveness character and mineral heterogeneity type. Heterogeneity management is under investigated methodologically however its role in ores processing is constantly growing. For example concentrate with 0.05% ferrous oxide was got from quarts with 1.5% of this component. This result could not be reached by other methods. It is important to provide transfer to homogeneous state when minerals have similar character of structural defects. Type and level of trybo treatment make it possible to solve these problems to certain extent.

3 Conclusions

Methodical bases of samples preparation in processes of mineral products extraction are reviewed. Taking into account orientation of these methods it is suggested to name this approach mineral preparation.

Possibilities of minerals separation by “Strat” method which is based on use of organic liquids with different density and surface trybo treatment are discussed. Mono mineral products of high quality are obtained as a result of methods application.

These methods as well as disintegrated minerals opening are recommended for wide application in analytical practice.

References

Berger GS (1962) Minerals’ floatating. Gortechizdat, Moscow
Ginzburg AI (1985) Mineral research methods Nedra, Moscow
Isotope dating of geological processes: new results, approaches and perspectives (2015) VI Russian conference in isotope geochronology. IPGG RAS, Saint Petersburg
Katz MY (1977) Minerals’ heterogeneity analysis. Nauka, Moscow
Methods and geo chronological results of isotope geometric minerals systems and ores studies (2018) VII Russian conference in isotope geochronology, Moscow
Yusupov TS, Baksheeva II, Rostovtsev VI (2015) Analysis of different-type mechanical effects on selectivity of mineral dissociation. J MiningSci 51:1248–1253
Yusupov TS, Travin AV, Yudin DS, Novikova SA, Shumskaya LG, Kirillova EA (2018) Improvement of methods of minerals – geo chronometers extraction from ores for geological processes isotope dating. In: Shchiptsov VV (ed.) Fundamental and applied aspects of technologica lmineralogy, Petrozavodsk. Russian conference in isotope geochronology, pp 86–92

25

Mineralogical-Geochemical Criteria for Geometallurgical Mapping of Levoberezhnoye Au Deposit

I. Anisimov(✉), A. Sagitova, M. Kharitonova, A. Dolotova, and I. Agapov

Polymetal Engineering JSC, Saint-Petersburg, Russia
anisimovis@polymetal.ru

Abstract. Levoberezhnoye gold deposit is located in Khabarovsk region. It formed quartz-sulfide and quartz-adularia veining and fracture zones in argillic altered intermediate volcanic tuffs and lavas. 61 variability study samples were composed of quartz, feldspar, mica, kaolinite, chlorite with minor pyrite, arsenopyrite, jarosite and accessories. Multivariate statistics of mineral composition and multi-element assays distinguished following ore types: (1) primary quartz-feldspar sulfide-bearing breccia veins, (2) oxidized breccias with micas transformed to illite-smectite; (3) high sulfidation quartz-kaolinite. Gold leach recovery correlated with high sulfate content as well as mica and chlorite transformation to illite and smectite. Low sulfidation ores showed lower leaching recovery connected to gold encapsulation in pyrite. Thus, oxidized and sulfate ore types were amenable to cyanidation, while primary ore was recommended for sulfide flotation gold recovery. Molybdenum high content connected to Ag, Cu, Pb and As and supposed to be formed in a separate mineralization event from gold.

Keywords: Gold recovery · Cyanidation · Oxidation · Illite-smectite

1 Introduction

Levoberezhnoye deposit is located in Khabarovsky region in Estern Russia. It is localized in intermediate volcanics and formed steeply dipping quartz-adularia Au-Ag breccia-vein system imbedded in rhyolites and extensively altered lake and flow tuffs and ignimbrite volcanics. The ore bearing rocks suffered multiple hydrothermal brecciation events with quartz±adularia-sulfide cement and fine sulfide dissemination in altered volcanics. The veining and rocks are fine grained and hard for visual mineral identification.

The samples characterized with drastic variations in gold recovery by cyanidation from 19 to 99%. The aim of the work was to determine compositional differences in ore types and ore characteristics effected gold recovery and cased metal losses.

2 Methods and Approaches

61 composite drill core sample of geotechnical mapping of Levoberezhnoe were studied for cyanadation leaching and bulk mineral composition. Sample color was described with RGB-parameters.

Au was assayed with fire assay with atomic absorption finish, multi-element ICP-AES assays after four acid digestion of straight and diluted samples and XRF-analysis, sulfide and total S, total C estimated by LECO analysis.

Mineral phase identification and their quantification was done using Eva software and COD database. Quantitative X-ray powder diffraction with Rietveld refinement Topas software at Polymetal Engineering.

Multivariate statistical analysis was performed on filtered data with Aitchison transformation using Pearson correlations with Cytoscape software, PCA and regression analyses.

3 Results and Discussion

Three main ore types were distinguished based on mineral composition: 1 – quartz-albite with sulfide (arsenopyrite-pyrite) mineralization, 2 – kaolinite-dickite with sulfate, 3 – illite-smectite.

Wide structural and chemical variety of feldspars was observed: high and low microcline, orthoclase, albite and hyalophane. Balancing mineral content QXRD results for micas and K-felpdspars revealed significant potassium shortage and suggested high baddingtonite content in feldspar, which needs confirmation with assays. Observed anomalously high values of molybdenum were connected to Ag, Pb, Au and Cu.

Multivariate statistics analyses included 73 following parameters: chemical and mineral composition, color (RGB, brightness – BRT and darkness - DRN), material fineness ($\gamma + 100$), Au recovery – εAu and losses – $-\varepsilon$Au. Pearson positive correlations revealed occurrence of 5 geochemical and mineral clusters (Fig. 1):

1 – quartz-kaolinite-albite-sulfate with zeolites, Sr enriched (high sulfidation alteration); 2 – micas-chlorite with calcite group was linked to 3 - sulfide mineralization (beresite – low sulfidation); 4 – potassic feldspar-ankerite; 5 - oxidation cluster between kaolinite and sulfide/micas.

Qu-Kln-Ab cluster showed kaolinitization. K-Fsp and mica clusters tied together with K and Rb. Barium feldspar - hyalophane associated with K-Fsp. Mica cluster combined muscovite, illite, illite-smectite and chlorite. Sulfide cluster connected with mica cluster and gold losses. Au recovery connected with As, illite, illite-smectite indirectly through the minerals formed from sulfide oxidation: jarosite, goethite and scorodite. Mica transformation to illite and illite-smectite followed oxidation of sulfides

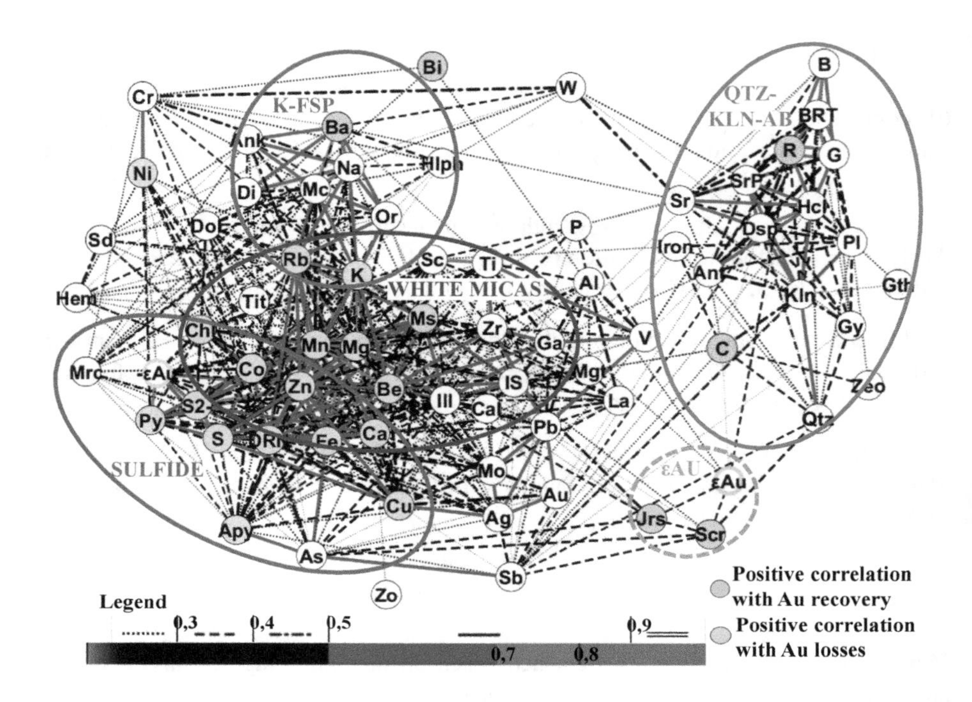

Fig. 1. Pearson correlations of 74 transformed chemical, mineral composition and cyanidation tests parameters of 51 small composite samples from Levoberezhnoye deposit

and, thus reflecting good Au leach results. Beresite group coupled with gold losses, the second ones indirectly correlated with gold recovery.

PCA analysis exposed 6 principal components, which explained 58.14% of the total variance. They described mineral composition (Fig. 2), oxidation rate, rare-metals and arsenic associations, color, grinding fineness. Results were similar to ones obtained with pair correlations: Au losses linked to sulfides, oxidation rate raised Au recovery (Fig. 2). Thus, flotation would be the best Au recovery solution from primary sulfide ore and tank cyanidation to oxide one.

Regression equation for gold recovery by cyanidation was calculated. It had relatively low $R^2 = 0.46$ (1) and connected color parameter (R/BRT) and elements contents (ppm and wt%):

$$\mathbf{\varepsilon AuCN} = -96.91 + \frac{185.3\mathrm{R}}{\mathrm{BRT}} - 7.83Ti(\%) - 2.71Co - 0.73La - 1.15Pb + 0.53Sc - 0.004Zr + 1.81Ga - 13.09Ssfd(\%) + 10.90Fe(\%) - 0.54V + 0.34Ag - 1.94Ni - 75.18Ca(\%) + 0.01Ba + 0.01P \quad (1)$$

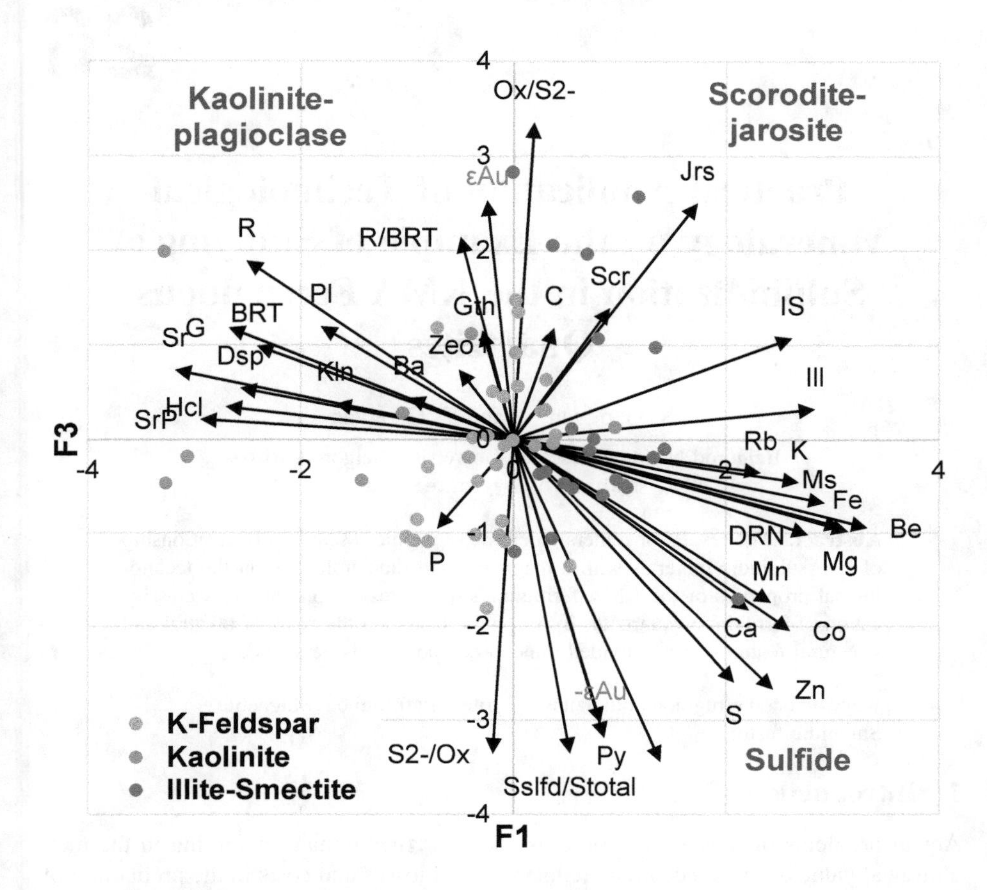

Fig. 2. Factor loadings and factor scores for factors 1 and 3 with interpretation

4 Conclusions

Mineral composition of the samples varied significantly from high quartz (up to 70%) and feldspar (up to 64%) to kaolinite (up to 45%) and illite-smectite (up to 40%). Au losses with cyanidation tails were bound to sulfides content. Sulfide oxidation to jarosite and scorodite and red component of sample color reflected increase in Au recovery. Absence of chlorite and transformation of muscovite to illite-illite-smectite also manifested Au cyanadation recovery improvement.

Revealed links between samples characteristics and regression equation, geometallurgical mapping and ore sorting can be performed based on sample color and chemical composition (Ti, Co, La, Pb, Sc, Zr, S, Fe, V, Ag, Ni, Ba).

26

Practical Application of Technological Mineralogy on the Example of Studying of Sulphidization in the KMA Ferruginous Quartzites

S. Gzogyan(✉) and T. Gzogyan

Belgorod National Research University, Belgorod, Russia
mehanobr1@yandex.ru

Abstract. The results of mineralogical-petrographic researches of relationship of iron sulphidic minerals with a magnetite and their influence on the technological properties of the KMA ferruginous quartzites are presented. A classification of ferruginous quartzites by sulphidic factor on the basis of textural and structural features of the divided mineral components is developed.

Keywords: Ferruginous quartzites · Pyrite · Pyrrhotine · Magnetite · Sulphidic factor

1 Introduction

Actual problems of processing iron ores and quartzites remain increasing in the metallurgical value of the concentrate, reducing metal losses and costs in the production of commercial products. In increasing the efficiency of processing ferruginous quartzites a special role belongs to the scientific direction – the technological mineralogy. The transition from the descriptive methodology to the methodology of genetic analysis allowed giving a scientific interpretation of many problems associated with the technology of processing ferruginous quartzites.

2 Methods and Approaches

The object of a research is more than 900 group geological and technological samples of the KMA ferruginous quartzites obtained during the operational exploration of deposits, which reflect the most representative texture-structural and mineralogical features of the KMA quartzite deposits.

The study of the composition and properties of the samples taken was carried out using optical microscopy, NMR-spectroscopy, high-temperature magnetometry, microprobe analysis and X-ray diffractometry.

The physical properties are used to determine the microhardness of minerals and magnetic properties. The main attention is paid to relationship of iron oxides and sulphides and their distribution in the finished product. Researches were carried out on samples of polished sections and briquettes.

3 Results and Discussion

The object of a research is more than 900 group geological and technological samples of the KMA ferruginous quartzites obtained during the operational exploration of deposits, which reflect the most representative texture-structural and mineralogical features of the KMA quartzite deposits.

The study of the composition and properties of the samples taken was carried out using optical microscopy, NMR-spectroscopy, high-temperature magnetometry, microprobe analysis and X-ray diffractometry.

The physical properties are used to determine the microhardness of minerals and magnetic properties. The main attention is paid to relationship of iron oxides and sulphides and their distribution in the finished product. Researches were carried out on samples of polished sections and briquettes.

The second category (medium enriched) includes ferruginous quartzites in which the structure of relationship between ore minerals and sulphidic minerals is more complex, that contributes to the transition to magnetite concentrate of both free grains of pyrrhotite and aggregates of magnetite with pyrite (Fig. 1b). From ferruginous quartzites of the second category (sulfur content 0.18), in the laboratory concentrate 0.09% remains (in the industrial one – 0.085%).

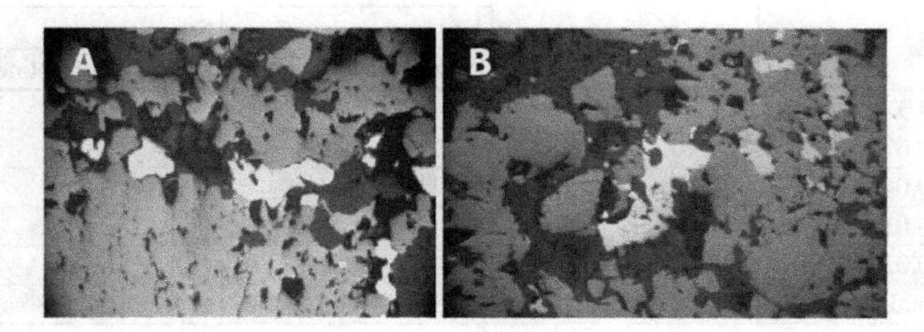

Fig. 1. Pyrite in magnetite aggregates (a) and internal and external intergrowths of pyrite with magnetite, poikilitic inclusions of pyrrhotite in magnetite (b), zoom 12.5 × 8 × 10.

The third category of quartzites (hardly enriched) is represented by ferruginous quartzites with close structural intergrowths of sulphidic minerals with magnetite, poikilitic inclusions of pyrite in magnetite and pyrrhotite and vice versa, and that leads to the transition and concentration of sulphides in the process of wet magnetic separation into magnetite concentrate (Fig. 2).

With a total sulfur content in the feedstock of 0.42–0.47 in the laboratory magnetite concentrate, its size reaches 0.42–0.78% (in the industrial one – 0.40–0.74%).

Thus, optical and mineralogical researches have shown that clogging of magnetite concentrate with sulfur occurs not only because of the ferromagnetic properties of the monoclinic pyrrhotite variety, but also due to the inclusion of pyrite grains in the magnetite and pyrrhotite grains and vice versa (Table 1).

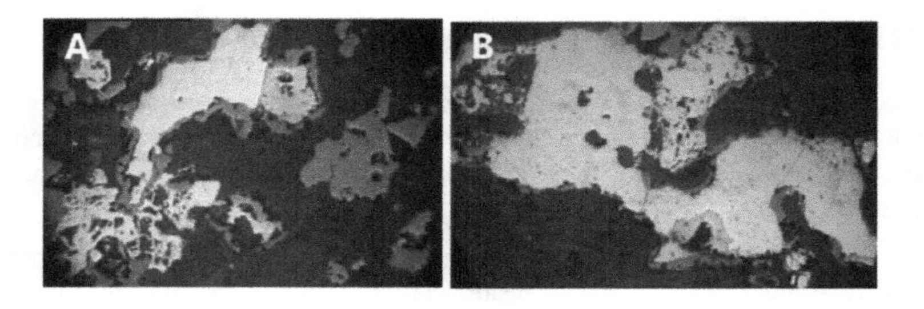

Fig. 2. Substitution structures of magnetite and pyrrhotite by pyrite, zoom 25 × 8 × 10 (a), 12.5 × 12.5 × 10 (b).

Optical and mineralogical researches of finished products have established that sulphidic minerals in concentrate are as in the form of intergrowths and inclusions in magnetite grains as in the form of separate grains (Fig. 3).

Table 1. Grain composition and distribution of chemical components by grain size in the magnetite concentrate

Grain size, mm	Yield of grain size,%	Content, %			
		Fe		S	
		Aggregate	Magnetite	Pyrite	Pyrrhotine
+0.071	1.20	29.4	27.53	0.046	0.040
−0.071 + 0.045	4.20	55.29	33.37	0.024	0.018
−0.045 + 0.032	27.33	66.92	64.88	0.028	0.019
−0.032 + 0.020	45.03	71.05	66.55	0.023	0.016
−0.020	22.24	70.47	62.77	0.021	0.014
Total	100.00	68.57	63.39	0.0242	0.0168

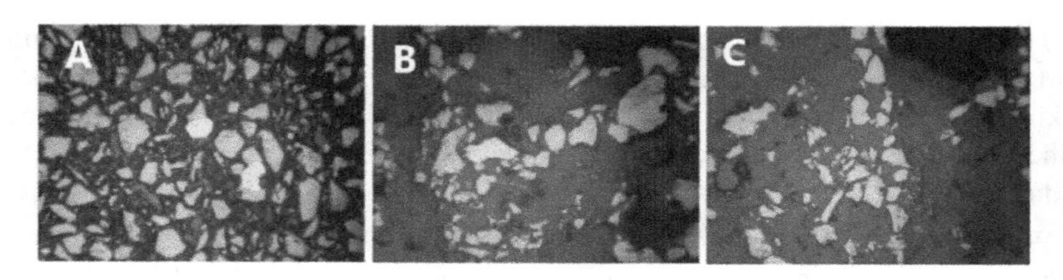

Fig. 3. Magnetite concentrate: aggregates of iron sulphides with magnetite and their separate grains (a, b) stand out against the background of free magnetite grains, poikilite inclusions in magnetite (c), polished briquette, zoom 132.

4 Conclusions

Thus, applying the methods of technological mineralogy and based on the studied nature of the variability of structural transformations, the classification of quartzites was developed according to enrichment by the sulphidic factor.

The next important achievement of technological mineralogy in the context of a particular sulphidic problem is the clearly shown migration of non-magnetic iron sulphides into magnetite concentrate using magnetic separation processes.

27

Process Mineralogy as a Basis of Molybdoscheelite Ore Preparation

L. Vaisberg[1(✉)], O. Kononov[2], and I. Ustinov[1]

[1] REC Mekhanobr-Tekhnica, St. Petersburg, Russia
gornyi@mtspb.com
[2] Moscow State University, Moscow, Russia

Abstract. The Tyrnyauz ore field (Big Tyrnyauz) is one of the largest and most geologically complex deposits of tungsten and molybdenum. The main valuable mineral of the deposit is molybdoscheelite, a representative of the sheelite—powellite isomorphic series, as well as molybdenite and respective accessory minerals. The main problem in the processing of ores of the Tyrnyauz ore field consists in the high variability of mineral associations of its rocks, including the availability of ores and nonmetallic minerals with similar physical and chemical properties, as well as in the availability of ores of various natural types. Selective mining and processing of various geological and industrial types of ores with a wide use of vibrational technologies for the selective disintegration of raw materials is a promising approach to the development of the Tyrnyauz concentrator.

Keywords: Molybdoscheelite · Tungsten · Molybdenum · Disintegration · Separation · Flotation

1 Introduction

The Tyrnyauz ore field is located in the North Caucasus and is one of the largest and most geologically complex deposits of tungsten, molybdenum and associated metals. The ore bodies of the Tyrnyauz deposit occur on the platform of Eljurta granites and diabases. In the central part, these have undergone intensive metamorphism and converted into amphibole-biotite hornfelses, overlain by numerous contact-metasomatic and hydrothermal formations. The main valuable minerals of the deposit are molybdoscheelite $Ca(W,Mo)O_4$, a representative of the isomorphous series of scheelite $CaWO_4$—powellite $CaMoO_4$, and molybdenite.

The deposit had been intensively developed starting from the middle of the XX century. Flotation had always been the main concentration process used at the plant. When the global market for tungsten changed at the end of the XX century, the Tyrnyauz concentrator was shutdown. By that time, the Main Skarn with tungsten and molybdenum mineralization and a significant part of the amphibole-biotite hornfelses with molybdenum mineralization had been mined out. At present, an upgrade of the Tyrnyauz concentrator is required. After the enterprise is repeatedly put into operation, its ore base will be represented by high-calcite skarn ore bodies (up to 10–15%) and skarned marbles with the calcite content of 40–70%. It is known that calcite and molybdoscheelite have similar flotation properties (Barskyi et al. 1979).

Due to the high calcite to molybdoscheelite ratios, their selective flotation with the use of traditional reagent regimes is impractical. Moreover, at lower mining horizons, scheelite is almost completely replaced by molybdoscheelite, its mineral variety with greater brittleness and lower hardness. These circumstances call for a new mineralogical and technological assessment of the Tyrnyauz deposit. Another independent promising type of ores are talcose hornfelses with high molybdenite contents and associated gold mineralization. However, the availability of a pair of naturally hydrophobic minerals (molybdenite and talc) also hinders their flotation concentration (Tarasevitch et al. 2014).

2 Methods and Approaches

Several hundred small and several dozens of large samples of promising ore types were selected and analyzed. Additionally, during the period of active operation of the deposit, the material composition and ore concentration indicators had been continuously analyzed for the respective production processes.

The grain-size distribution of molybdoscheelite (MSh) of different compositions and powellite is due to the regular decrease in hardness and increase in brittleness with the increase in the molybdenum content in the scheelite—powellite series. The higher hydrophobicity of powellite and high-molybdenum MSh (with powellite inclusions), which increases for particles of fine classes due greater specific surface values, predetermines their accumulation in the sulfide-molybdenum concentrate. In contrast, the higher affinity between scheelite and fatty acids, as compared to MSh and powellite, gives certain advantages in the process of rougher scheelite flotation and in respective final treatment.

The dependence of flotation properties of MSh in these operations on the composition, chemical and phase inhomogeneity of the particles enables the use of MSh composition and dispersion values to evaluate the efficiency of flotation processes.

Chemical sampling data on the contents of Mook and WO3, their ratio and related dispersion and the results of MSh local composition analysis serve as the criteria for assessing the chemical inhomogeneity of MSh. The color of luminescence, spectroscopic and kinetic parameters of excitation and radiation represent rapid test indicators for MSh. The color of luminescence is effective when assessing the content of pure scheelite with high hardness and flotability (blue luminescence) and of all varieties of MSh with powellite characterized by low hardness and poor flotability (yellow luminescence). When predicting the processing properties of MSh, dispersion assessment data is of greater significance. Therefore, the criteria used for identifying the processing varieties of ores with account of the composition of MSh should include variations in the composition of pure scheelite, molybdoscheelite and powellite. Similar distribution of molybdenite oxidation values in the ore biotite hornfelses (with almost zero MSh content) occurring in parallel to the Slepaya deposit may be used to assess the content and distribution of powellite.

Skarns of the Slepaya deposit do not form independent bodies. These either form part of the metasomatically altered skarned hornfelses, layered and massive marbles, or are represented by relics in the areas of development of later metasomatic mineral associations. In both cases, the skarns are represented by thin lens-shaped and clustered units (with the thickness of several centimeters to several tens of centimeters) having a fine to medium-grained structure with the characteristic light greenish-brown color.

Marbles in the skarn bodies are white, light gray and dark gray massive rocks with a spotted, brecciated or banded texture caused by the alternating interlayers of dark and light marbles of different granularity, alternating interlayers of light marbles and thin banded light green pyroxene-plagioclase hornfelses and new units in the form of wollastonite-fluorite, pyroxene-fluorite and quartz-pyroxene-calcite-garnet veins, veinlets and lenses. The marbles are composed of calcite. The sizes of polysynthetic calcite grains containing fine scattered inclusions of graphite and pyrrhotite range from 0.08 to 1.6 mm.

In the skarned marbles, the distribution of structural, textural and color varieties of marble is non-uniform. Within the contours of the ore body, fine, medium and non-uniformly crystalline and, respectively, dark and light sulfur varieties predominate (in quantitative terms). The lighter and white varieties ("clarified" marbles) form among them a system of veinlets, lens-shaped interlayers, clusters and blocks and are part of the parallel-banded varieties of marbles with calcite bands with a thickness of 1 to 5 cm. Their formation is associated with recrystallization along the fractures and increased permeability zones, as well as in tectonic deformation processes. In addition, lighter zones are always observed at the interfaces between marbles and skarns and around quartz-silicate veins and veinlets.

Luminescent properties of the ore minerals and host rocks of the Tyrnyauz deposit were studied since it is possible to use these luminescent properties as indicators for rapid diagnostics and as separation features in ore separation.

Preliminary concentration of the Tyrnyauz ore, which may be implemented using X-ray luminescent separation (RL), is required due to the processing of comparatively low-grade ores with large amounts of diluting materials and high mineralization contrasts. The contrasts in the luminescent properties of minerals and, therefore, of rocks and ores, may also be used for the rapid diagnostics of various natural and industrial types of ores in the process of their transportation and sorting, for optimizing their grinding conditions, as well as in their subsequent processing. The RL spectra were registered in the continuous and staged scanning modes. In the continuous scanning mode, with the constantly enabled source of excitation, the entire spectrum was recorded. This is the traditional and most informative form of data recording. However, under the influence of X-rays, the luminescence intensity of a substance usually increases. After a certain period of time that is specific for respective centers of luminescence, the saturation level is reached, when the luminescence intensity practically ceases to change or changes so slowly that it produces no effect on the spectrum image at any moment of the scanning time.

Therefore, reproducible and comparable results in the continuous scanning mode may only be obtained when the samples are irradiated to the saturation state, i.e. for relatively long periods (tens of minutes or more, depending on the mineral). In this connection, a method was developed for expressly obtaining the required information on the spectra by means of high-speed analysis in pulsed mode with staged scanning of the spectrum. Spectral and kinetic RL characteristics were studied for the following minerals: scheelite and molybdenite, calcite, wollastonite, quartz and plagioclase. The resulting RL spectra were interpreted by comparison of the data obtained with the reference data. For all minerals, respective types of luminescence centers were determined and the issues of variability in luminescent properties were considered. For the spectra obtained, the intensities, positions of the maximum and half-widths of the intrinsic luminescence band associated with the [WO4] and [MoO4] groups were established, as well as the wavelength and the intensity of the luminescence lines for impurity ions of rare-earth elements isomorphically replacing calcium ions in the scheelite structure.

3 Results and Discussion

Visual observation of the luminescence for the molybdoscheelite samples studied showed that many of them are characterized by a non-uniform color glow (blue and yellow), which indicates joint availability of several varieties of scheelite. In this case, the experimentally studied RL spectrum is composed of individual spectra of these varieties and represents an averaged characteristic for the scheelites of the sample. The scheelite and molybdoscheelite samples studied are characterized by varying degrees of chemical composition inhomogeneity, which is manifested in different colors of their fluorescent luminescence. The half-width of the band of intrinsic luminescence may be used as the inhomogeneity measure.

The observed diversity of the spectra may be explained by the presence of different types of MSh in different sample groups with varying content ratios and by the presence of their spectra. Based on the spectroscopic data obtained, at least four varieties with different luminescent properties may be distinguished for the scheelite samples studied. Signs of vertical zoning are observed in the distribution of MSh with different types of luminescence spectra. At the uppermost horizons, the luminescence intensity is usually higher.

The luminescence and afterglow spectra for all calcite samples studied consist of a single broad band with the luminescence maximum of 630 nm (orange-red). It was also found that luminescence intensity is variable and depends on a number of parameters, such as the duration of x-ray irradiation, the concentration of manganese, and the color of the samples. An important element of the promising processing technology for the Tyrnyauz ores is the use of selective vibration in disintegration and classification of raw materials.

4 Conclusions

These studies in the field of process mineralogy of molybdoscheelite ores of the Tyrnyauz deposit form the basis for the design of a new combined process flow using various selective vibration methods for disintegration and preliminary lump separation.

Acknowledgements. The study was supported by the grant of the Russian Science Foundation (project No. 17-79-30056).

References

Barskyi LA, Kononov OV, Ratmirova LI (1979) Selective flotation of calcium minerals. Nedra, Moscow

Tarasevich Y, Aksenenko EV (2014) The hydrophobicity of the basal surface of talc. Colloid J 76(4):526–532

28

Ore Mineralogy of Kirazliyayla Mesothermal Zn-Pb-(±Cu) Deposit

F. Javid and E. Çiftçi(✉)

Department of Geological Engineering, Faculty of Mines, ITU, 33469 Maslak, Istanbul, Turkey
eciftci@itu.edu.tr

Abstract. The Kirazlıyayla deposit is one of Pb-Zn (±Cu) deposits associated with andesitic volcanism cutting through the metamophics of Karakaya complex. It is a structurally controlled, vein-stockwork style, mesothermal ore deposit. Replacement, brecciation, vein/veinlets, carries, sea-island, and dissemination textures were identified. Pyrite, sphalerite, chalcopyrite, galena, tennantite, and covellite constitute the ore mineral paragenesis. Quartz, calcite and dolomite with kaolinite account for the gangue minerals. Supergene stage is insignificant. The Kirazlıyayla mineralization is a Zn-Pb (±Cu) mineralization hosted by tectonically controlled andesitic volcanism within the Karakaya metamorphics and has common features with the other occurrences in the western Anatolia.

Keywords: Kirazlıyayla · Mesothermal · Lead · Zinc · Karakaya complex · Structurally controlled

1 Introduction

Turkey hosts noteworthy variety of mineral deposits owing to its geological evolution within the Alpine-Himalayan orogenic belt and its complex tectonic setting. Mineral deposits of Turkey are better understood through the understanding tectono-magmatic evolution of Turkey. Within this framework, there are a number of various ore deposits disseminated throughout the country including the Kuroko-type VMS deposits (strictly in the Eastern Pontides tectonic belt), Cyprus-type VMS deposits (along the Bitlis-Zagros suture zone in SE Anatolia and Küre-Kargı trend in the Central Pontides), Besshi-type VMS deposits (in the Central Pontides and along the Bitlis-Zagros suture zone), epithermal deposits of both LS and HS type (mostly in western Anatolia) and significant number of IS type with noteworthy lead and zinc presence, and carbonate-hosted sulfidic and nonsulfidic Pb-Zn deposits (along the Tauride Belt). In addition, skarn-type (mainly Fe producing) deposits (disseminated throughout the country), porphyry Cu-Mo (disseminated throughout the country), mesothermal cupriferous Pb-Zn mineralizations (along the Eastern Pontide belt), and podiform chromite deposits (along the suture zones throughout the country) are also important part of the metallogeny of Turkey.

Although Turkey produces zinc and lead concentrates, due to the lack of smelters, about 1.5 billion dollars each year paid for the import of those metals. In the recent years, the government is in an attempt to revive the sector through various incentives.

The western Anatolia is host to great number of mineral deposits. As for the lead-zinc occurrences, it can be summarized in 3 types in terms of their genesis in the region: (I) distal skarn occurrences associated with the carbonates of the metamorphic basement; (II) structurally controlled veins within the basement metamorphics and the overlying volcanics or along their contacts; and (III) replacements along the mafic dykes.

The study area is located in one of the tectonic elements of Turkey – the Sakarya terrane, which is an elongate crustal ribbon extending from the Aegean in the west to the Eastern Pontides in the east. It is consisted of sandstones of Lower Jurassic age, which sits on a fairly complex metamorphic basement that contains a high-grade Variscan basement metamorphics of Carboniferous age (Topuz et al. 2004, 2007; Okay et al. 2006), Paleozoic granitoids (Delaloye and Bingöl 2000; Okay et al. 2002, 2006; Topuz et al. 2007), and a low grade metamorphic complex - the Lower Karakaya Complex constituted by Permo-Triassic metabasite with lesser amounts of marble and phyllite. The Lower Karakaya Complex represents the Permo-Triassic subduction-accretion complex of the Paleo-Tethys as indicated by the presence Late Triassic blueschists and eclogites (Okay and Monié 1997; Okay et al. 2002), accreted to the margin of Laurussia during the Late Permian to Triassic. The complex is overlain by a thick series of strongly deformed clastic and volcanic rocks with exotic blocks of Carboniferous and Permian limestone and radiolarian chert. This complex basement was overlain unconformably in the Early Jurassic by a sedimentary and volcanic succession. The Early Jurassic is represented by fluvial to shallow marine sandstone, shale and conglomerate in the western part of the Sakarya Zone. The metamorphic basement is cut by Eocene volcanisms.

The Kirazliyayla Zn-Pb ore deposit is spatially and temporally related with Eocene intermediate extrusive rocks - andesite and trachyandesite with NE-SW extension and covered by clastic and carbonate rocks. The main purpose of the investigation is to determine the genesis of Kirazliyayla Zn-Pb ore deposit and its place in the metallogenic evolution of the region to contribute the understand of the metallogeny of Turkey. In that, geochemical characteristics, mineralogy of both host rocks and ore minerals, its tectonic setting are the main issues are covered.

2 Methods and Approaches

A total of 30 samples representing the ore deposit from open pit – main production step and from the boreholes were taken, from which polished sections were prepared and Nikon Eclipse LV100 reflected light microscopy integrated with a CITL MK5 Cathodoluminescence system was employed for mineralogical examination. The ore minerals and the paragenesis were identified on the basis of their petrographical features and their textural relationships, respectively. Electron Probe Micro Analysis (EPMA) and Secondary Electron Microscopy-Energy Dispersive Spectroscopy (SEM-EDS) were routinely used for confirming the minerals and chemistry of sulfide minerals when needed.

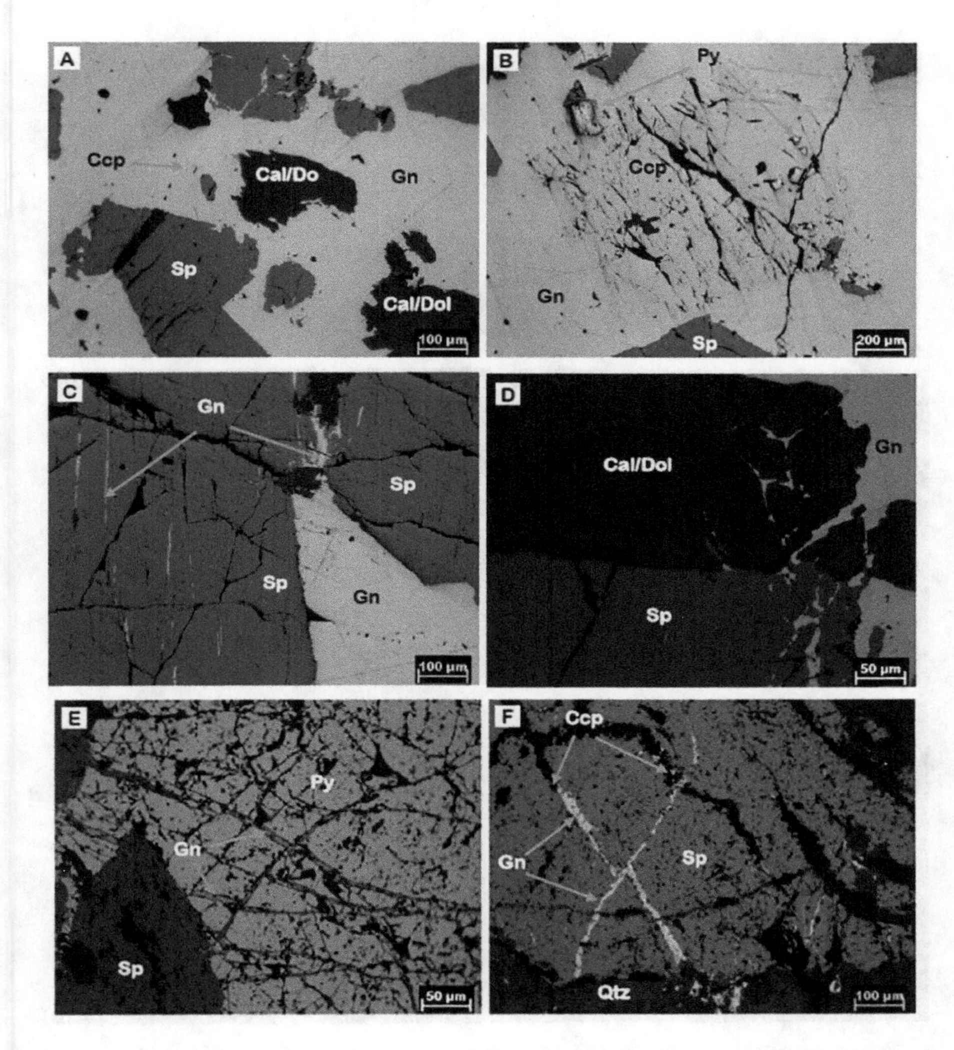

Fig. 1. (A) sphalerites and carbonates are partially replaced by late stage galena; (B) chalcopyrite and pyrite are partially replaced by late stage galena; (C) galena replaces coarse sphalerite; (D) galena replaces coarse sphalerite and also carbonate gangue; (E) large cataclastic pyrite veined by sphalerite and late stage galena; (F) large cataclastic sphalerite with replacing galena and chalcopyrite, all in silicic matrix; (G) Large sphalerite grains with late stage galena, chalcopyrite and tennantite veins; (H) large pyrite and chalcopyrite veined by late stage galena; (I) Euhedral quartz, large sphalerite grains with late stage galena; (J) sphalerite veined by late stage chalcopyrite and pyrite; (K-L) Late carbonates veining sphalerite (Py: pyrite; Ccp: chalcopyrite; Sp: sphalerite; Gn: galena; Tn: tennantite; Cal/Dol: calcite/dolomite)

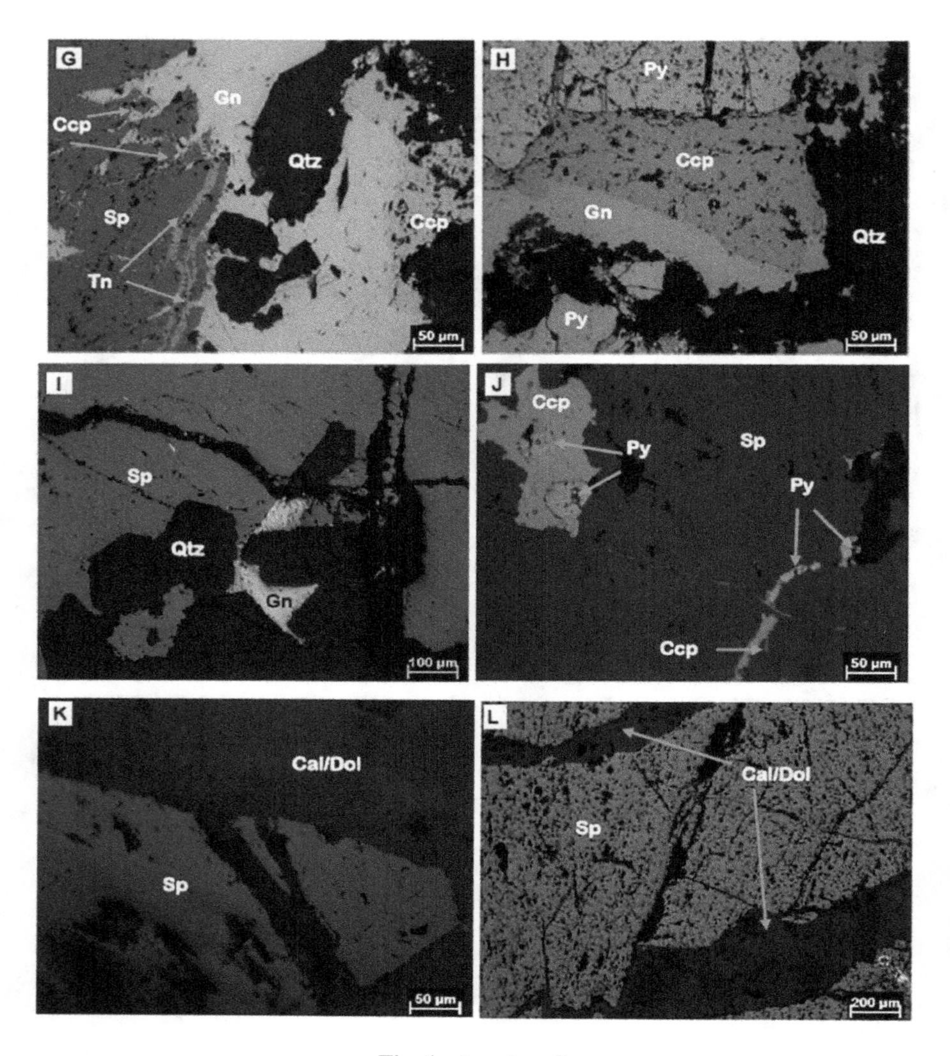

Fig. 1. *(continued)*

3 Results and Discussion

Field observations indicated that kaolinitization with silicification is pervasive alterations indicating low pH hydrothermal activity resulted in the mineralization. The mineralization is associated with andesitic volcanics in that vein/veinlets, replacement in places, stockwork, and breccia ore structures are prevalent. Thickness of veins may reach to 1 m in rare occasions. Samples were collected from the production steps in the open pit and from the boreholes cutting the mineralized zones.

Based on the studies of the ore samples, the major ore minerals are sphalerite and galena as zinc and lead carrier. Pyrite is a ubiquitous. Chalcopyrite is a minor to trace sulfide phase along with trace tennantite (Fig. 1K–L). Most of the samples indicate that the mineralization is medium to high grade. Sphalerite and galena occur as large grains (sometimes up to mm size) in most of the samples, indicative of being precipitated out of supersaturated fluids within narrow spaces. Both are also fairly inclusion-free (clean). When not, galena occurs with tennantite, sphalerite with pyrite and chalcopyrite. Galena has also chalcopyrite encapsulation in places.

Interpretation of the intergrowth ore textures suggests that mineralization started with pyrite crystallization then a brief precipitation of first generation of chalcopyrite, then followed by major sphalerite crystallization, which is followed by a major carbonatization, followed by a brief tennantite and second generation of chalcopyrite formation. In the final stage of the ore mineralization a major galena precipitation took place. Figure 2 summarizes the mineralization event. Ore deposit experienced very weak supergene stage in which only traces of covellite formed.

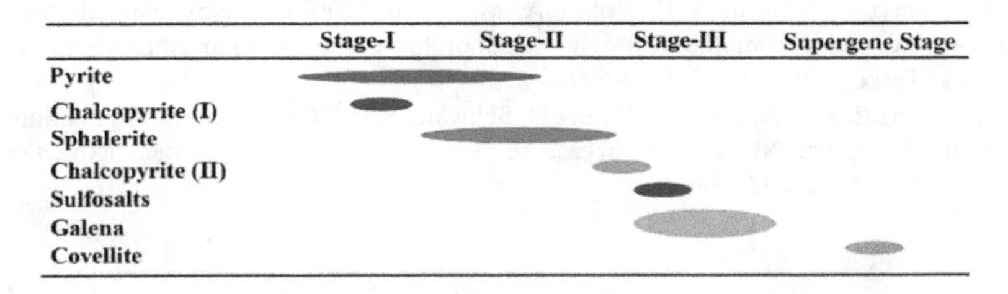

Fig. 2. Proposed ore mineral paragenetic sequence for the Kirazlıyayla Zn-Pb mineralization

4 Conclusions

Major ore minerals include sphalerite and galena. Chalcopyrite is minor while tennantite is trace. Pyrite is always present, but not as much as sphalerite and/or galena. Tennantite is the only fahlerz that occurs in some ore zones, overall in trace quantities. Sphalerite and galena occur as large grains in most of the samples. Both are also fairly inclusion-free (clean). When not, galena occurs with tennantite, sphalerite with pyrite and chalcopyrite. Galena has also chalcopyrite encapsulation in places. Paragenetic succession for the ore minerals appear to be (from early to late):

Pyrite-Chalcopyrite (I)-Sphalerite-Chalcopyrite (II)-Tennantite-Galena

Calcite/dolomite and quartz account for the gangue minerals. Kaolinite is the major clay mineral. Sericite also locally become significant.

Coarse nature of the major minerals (majority of the galena and sphalerite is larger than 100 microns) suggests high liberation (>90%). Small size occurrences (<10%) are mostly in the form of veinlets within each other, so during milling, most of that size range may be liberated.

Sulfosalts or fahlore are represented only by tennantite ($Cu_6[Cu_4(Fe,Zn)_2]As_4S_{13}$).

Preliminary results show that the Kirazlıyayla mineralization is a Zn-Pb (±Cu) hosted by andesitic volcanics whose emplacement within the metamorphic complex should be tectonically controlled.

Acknowledgements. The authors would like to acknowledge support received from Meyra Mining company for their courteous support during the field work.

References

Delaloye M, Bingöl E (2000) Granitoids from western and northwestern anatolia: geochemistry and modeling of geodynamic evolution. Int Geol Rev 42:241–268

Okay AI, Tüysüz O, Satır M, Özkan-Altıner S, Altıner D, Sherlock S, Eren RH (2006) Cretaceous and Triassic subduction-accretion, HP/ LT metamorphism and continental growth in the Central Pontides, Turkey. Geol Soc Am Bull 118:1247–1269

Okay Aİ, Monié P (1997) Early Mesozoic subduction in the Eastern Mediterranean: evidence from Triassic eclogite in northwest Turkey. Geology 25:595–598

Okay Aİ, Monod O, Monié P (2002) Triassic blueschists and eclogites from northwest Turkey: vestiges of the Paleo-Tethyan subduction. Lithos 64:155–178

Topuz G, Altherr R, Schwarz WH, Dokuz A, Meyer HP (2007) Variscan amphibolite facies metamorphic rocks from the Kurtoğlu metamorphic complex (Gümüşhane area, Eastern Pontides, Turkey). Int J Earth Sci 96:861–873

Topuz G, Altherr R, Kalt A, Satır M, Werner O, Schwartz WH (2004) Aluminous granulites from the Pulur Complex, NE Turkey: a case of partial melting, efficient melt extraction and crystallisation. Lithos 72:183–207

29

Geochemistry and Mineralogy of Copper Mine Tailings

K. Berkh(✉), D. Rammlmair, M. Drobe, and J. Meima
Federal Institute for Geosciences and Natural Resources, Hanover, Germany
Khulan.berkh@bgr.de

Abstract. Selected mine tailings in northern-central Chile were geochemically and mineralogically studied for their economic potential and environmental impact. High bulk Co content up to 1500 ppm and Cu content up to 9100 ppm are caused by Co-bearing pyrite, chalcopyrite, and their secondary products such as malachite and Co-Cu-carrying Fe-hydroxides. Due to high amount of sulfide minerals acid mine drainage (AMD) is forming in the oxidized upper part of the tailing, which makes a retreatment in dispensable to reduce the environmental impact.

Keywords: Mine tailings · Geochemistry · Mineralogy · Cobalt · Copper · Reprocessing

1 Introduction

Since Chile is the largest Cu producing country with the biggest reserves, Chilean mining industry generates huge quantities of mining residues, amongst others in the form of tailings dumps. The fact that some of them were generated many decades ago, where process technology was inadequate in comparison to today, makes some of them economically interesting. Due to advanced weathering, the tailings, which are potentially acid producing, bear an environmental hazard. Therefore, our aim is to investigate the geochemistry and mineralogy of the tailings in order to determine their economic potential and environmental impact.

The studied mine tailings dump is located in the region of Coquimbo, where an arid Mediterranean climate with mean annual temperature of 14.6 °C, precipitation of 132 mm, and evaporation of about 1702 mm prevails (Mora et al. 2007). Geologically, the region is characterized by Early Cretaceous Chilean Manto-type (volcanic-hosted stratiform) Cu deposits (Kojima et al. 2003) that are the most probable source of the studied tailings.

2 Methods and Approaches

Eight drill cores with lengths of seven meters were taken from the tailings dump (Fig. 1). Bulk geochemistry of homogenized material for each meter was investigated using a standard WDXRF. Mineralogy was studied on representative grain concentrates obtained by gravity separation. Abundance of the minerals was analyzed by MLA and composition of the minerals was examined by EPMA.

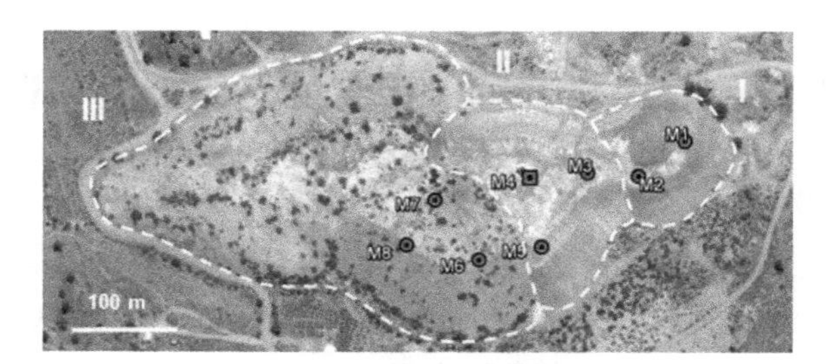

Fig. 1. Drill cores positions (M1–M8) in three heaps (I-III)

3 Results and Discussion

The tailings material consists of alternating layers of sand, silt and clay. According to geochemical pattern, the tailings dump can be subdivided into three groups, as shown in Fig. 1. **Heap I and II** are enriched in Fe and Co (Fig. 2). Special feature of heap I is a depletion in S and Ca but a high percentage of loss on ignition (LOI). It may point to a weathered part of the tailings dump, where oxidation of pyrite results in dissolution of calcite and accumulation of water bearing secondary clay minerals. **Heap III** contains high amounts of host rock and therefore elevated contents of Si, Al, Mg, Na, K and P can be seen. It is strongly enriched in Cu but only at the near surface level (Fig. 2).

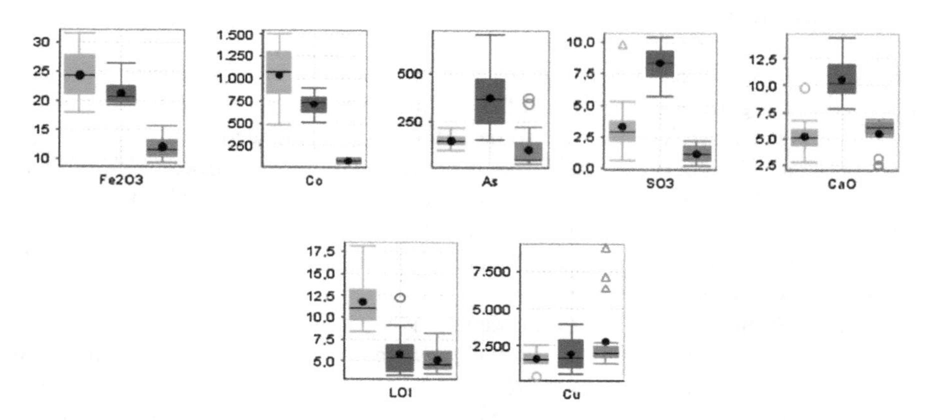

Fig. 2. Bulk content of relevant elements in three heaps (▪ – heap I, ▪ – heap II, ▪ – heap III)

Primary gangue minerals in **heap I and II** are epidote, quartz, andradite, and albite pointing towards skarn mineralization. The only difference between the two heaps is the absence of calcite in the heap I confirming the bulk geochemistry. The most common primary ore mineral in both heaps is pyrite, which occurs as liberated grains. Three types of pyrite were identified. The first type is pure pyrite. The second type has an As-rich rim with up to 8.7 wt% As (Fig. 3a) and is predominantly present in the heap II resulting in an elevated bulk content of As (Fig. 1). The third type contains a significant amount of Co. Hereby, the Co concentration increases from rim to core and can reach up to 3.8 wt %. Additional trace elements are e.g. up to 0.5 wt% Ni, 0.1 wt% Cu, and 0.2 wt% Cd. Remaining primary ore minerals are magnetite, hematite and trace amounts of chalcopyrite. They also occur as well liberated grains and are occasionally inter grown with each other. These minerals do not contain significant amounts of trace elements.

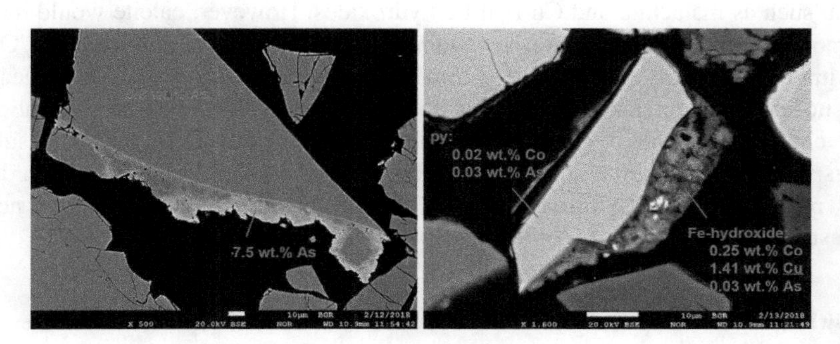

Fig. 3. BSE images of the pyrite grains: a As-rich rim on pyrite and b pyrite rimmed by Fe-hydroxide with significant amounts of Co and Cu

Secondary phases are gypsum and Fe-hydroxides as weathering products of calcite and pyrite (Fig. 3b) and preferentially occur in heap I. Such in-situ precipitation of Fe-hydroxides can only take place at near neutral pH proving an initial occurrence of calcite as a buffering agent in the strongly weathered heap I. The Fe-hydroxides contain high concentrations of Co (up to 0.9 wt%), Cu (up to 2.5 wt%) and Ca (up to 4.3 wt%), pointing to a dissolution of pyrite, chalcopyrite and calcite and fixation of released metals in Fe-hydroxides. The analyzed saturated soil extraction had a pH of 2. Its Co and Cu contents were 103 and 136 mg/l respectively, whereas Hg, Cd, As, Zn and Pb contents remained under 2.4 mg/l. Especially, the Co and Cu values exceed the international guidelines (IFC 2007 and Gusek and Figueroa 2009) by more than a factor of 100.In fact, the potential for AMD is high in the oxidized parts of the heaps.

The **heap III** predominantly consists of gangue minerals. The most common mineral is quartz followed by albite, anorthite, micas, K-feldspar, amphibole, epidote, pyroxene, chloride, and calcite. A Minor amount of Fe-oxides and chalcopyrite exists, either as liberated grains or finely intergrown with gangue minerals. Chalcopyrite is strongly replaced by Fe-hydroxides and Cu-carbonates along grain boundaries and micro fractures.

4 Conclusion

Bulk geochemistry provides a good prediction of the mineralogy. Elevated amounts of Fe, Co, and S in the heaps I and II correspond to Co-bearing pyrite-rich waste material. In contrast, heap III consists of host rock forming elements resulted by non-sulfidic gangue mineral waste. From an economic point of view, Co is the only valuable metal in the heap I and II. For the extraction magnetic separation should be performed to eliminate high amounts of magnetite. Heap I contains extensive amounts of Co- and Cu-rich Fe-hydroxides that usually coat pyrite grains. Therefore, leaching can be directly applied to extract acid soluble Co and Cu and also to liberate pyrite grains. Otherwise, pyrite cannot be floated. Afterwards, pyrite from both sulfidic heaps can be floated and Co can be extracted by bioleaching. In case of heap III, only the uppermost first meter, which is Cu-rich, should be treated. Cu occurs preferentially as acid soluble mineral such as malachite and Cu-rich Fe-hydroxides. However, calcite would have to be removed by gravity separation to reduce the consumption of sulfuric acid. Otherwise, itis economically not feasible. From an environmental point of view, heap III poses no environmental risk because it does not host AMD potential. Oxidation of minor chalcopyrite will be buffered by the carbonate content. In turn, the sulfidic tailings should be immediately treated, because the potential of AMD generation due to heavy rains is high. The hazardous potential of the impoundment should not be underestimated because of agricultural activities in the vicinity.

Acknowledgements. This investigation is supported by DERA the German Mineral Resources Agency of BGR.

References

Gusek JJ, Figueroa LA (2009) Mitigation of metal mining influenced water (Management Technologies for Metal Mining Influenced Water, vol 2), Soc Min Metall Explor. Littleton, Colorado

IFC (2007) Environmental, health and safety guidelines for mining. International Finance Corporation, World Bank Group, DC, 33 p

Kojima S, Astudillo J, Rojo J, Tristá D, Hayashi K (2003) Ore mineralogy, fluid inclusion, and stable isotopic characteristics of stratiform copper deposits in the coastal Cordillera of Northern Chile. Miner Deposita 38:208–216

Mora F, Tapia F, Scapim SA, Martins EN (2007) Vegetative growth and early production of six olive cultivars, in Southern Atacama Desert, Chile. J Cent Eur Agric 8(3):269–276

30

Use of Borogypsum as Secondary Raw

A. Khatkova[1], L. Nikitina[2], and S. Pateyuk[2](✉)

[1] Department of Mineral Technology, School of Geology, Transbaikal State University, Chita, Russia

[2] Department of Geology, University of Kazan, Kazan, Russia

nesvvik@gmail.com

Abstract. We have considered problem of accumulation, storage, utilization and recycling of wastes of various industries. We chose borogypsum as the object of study, which contains gypsum and silicon dioxide, which can be used in various industries. We proposed a new flotation reagent for separation of silicon concentrate from wastes of boric acid production. Using methods of mathematical planning we conducted a multifactorial experiment, which allowed identifying optimal flotation mode. We developed a technology for processing of borogypsum.

Keywords: Secondary raw · Borogypsum · Flotation · Perlastane · White soot · Silicon concentrate

1 Introduction

Over the years the enterprises of mining and chemical industries in the Far East accumulated millions tons of industrial wastes that are currently not being recycled. Thus, the total amount of wastes from the production of boric acid - borogypsum - in the Far Eastern region is more than 25 million tons. Borogypsum contains gypsum and silicon dioxide, which can be used in various industries. In this regard, the problem of complex processing of these wastes to obtain various functional materials is a very urgent task (Gordienko et al. 2014).

Highly dispersed amorphous silica is referred to as white soot. White soot of all kinds, alongside with black ones, is used to strengthen rubber, being absolutely indispensable for silicone rubbers. It is also used as fillers in the production of rubber linoleum, as well as in plastics, paints, lubricants and other materials to give them valuable properties.

In addition to silicon dioxide, another component of borogypsum, gypsum and anhydrite, can be used in industry. All possible products (component for cement production, dry building mixtures, finished building and architectural products, gypsum boards) have positive market prospects in Russian markets and in some neighboring foreign markets.

In Russia there are projects for the production of white soot with a high silica content by separating from datolite tailings. However, such technologies are not implemented at industrial scale, because they are quite expensive due to the low content of silicon dioxide in the original product.

To reduce the consumption of acids and the cost of production of white soot, it is necessary to increase the content of silicon dioxide in the original product by other cheaper methods such as flotation.

The table below shows the dependence of the need for concentrate on the content of silicon dioxide (Table 1).

Table 1. The dependence of the need for concentrate on the content of SiO2

SiO_2 content in concentrate	Need for concentrate, t/hour	Need for sludge t/hour, with a concentrate yield 15%	Need for pulp volume, m^3/h, at 20% solids
65	1,35	9	40,1
60	1,47	9,8	43,7
55	1,60	10,7	47,7
50	1,76	11,7	52,1

According to the table, it is necessary to ensure the content of silicon dioxide from 50 to 65% with a product yield 15% and above using flotation.

2 Methods and Approaches

A sample of waste of boric acid production, borogypsum, obtained from Dalnegorsky MPP was selected as the object of study.

For the flotation of gypsum-containing raw, fatty acid reagents are commonly used (Bulut et al. 2008; Matsuno et al. 1958). We proposed a new flotation reagent perlastan ON 60 to isolate the silicon concentrate by reverse flotation. It was efficiently used at the flotation of non-metallic fluorite ores (Dolgikh 2012), and is a promising flotation reagent for borogypsum.

To determine optimal conditions for flotation, we used the method of rational planning of a multifactorial experiment (Malyshev 1977).

3 Results and Discussion

After conducting a multifactorial experiment, the following regularities were identified.

Increasing temperature of flotation leads to decreasing extraction, but at the same time, the content of silicon dioxide increases. It was also found that the temperature affects the floatability of gypsum, however, almost no effect on the floatability of silicon concentrate. Consequently, by changing the temperature, it is possible to regulate the process, depending on the task. Changes in flotation pH slightly affect the content and recovery, therefore, pH control is not effective, which in turn is an advantage of this method, for it does not require consumption of additional reagents. Increasing concentration of perlastane increases the content of silicon concentrate in the chamber product, however, the extraction drops significantly. The agitation time has a linear effect on the extraction, at the same time we didn't observe a clear dependence of the agitation time on the content. Therefore it is necessary to agitate perlastane with minerals for at least 3 min. As the flotation time increases, the content slightly decreases due to the fact that alongside with calcium sulphate, silica dioxide begins to shift to the foam product.

On the basis of the revealed regularities we determined optimal conditions of flotation. However, the content of silicon concentrate in the chamber product remains insufficient.

From the literature it was found that sodium sulfide or liquid glass was often used to suppress flotation of quartz. Experiments with these reagents showed that they practically did not suppress flotation of amorphous silicon dioxide, and could also suppress flotation of gypsum, therefore it would be not advisable to use them for borogypsum flotation.

We determined that it would be much more effective to increase the concentration of perlastane at the main flotation, increasing its time than to feed the reagent fractionally in different operations. Increasing the concentration of perlastane up to 1 kg/t, as well as flotation time up to 9 min, allows obtaining a product with a content of 53.6%, with a yield 32.5% and a recovery 64%.

4 Conclusions

We suggested a new technology for processing wastes of boric acid production according to the scheme presented in Fig. 1. The proposed scheme has its advantages in comparison to previously known technologies. Thus it does not require regulation of pH of the medium, consists of a single operation, which significantly reduces flotation time and also allows to increase the yield of the chamber product significantly.

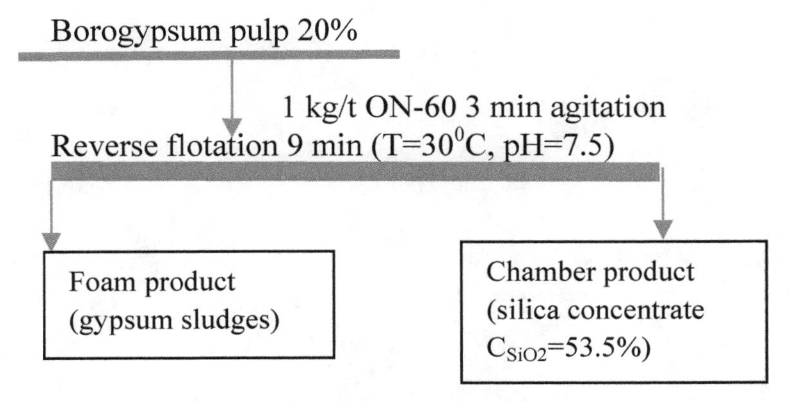

Fig. 1. Diagram of borogypsum flotation

References

Bulut G, Atak S, Tuncer E (2008) Celestite-gypsum separation by flotation. Can J Metall Mater Sci 47(2):119–126

Dolgikh OL (2012) The use of perlastane reagent as an alternative to oleic acid during fluorite flotation. Bull ZabGU 9(88):20–26

Gordienko PS, Kozin AV, Yarusova, SB, Zgibly IG (2014) Integrated processing of waste from the production of boric acid to produce materials for the construction industry. Min Inf Anal Bull (S4-9):60–66

Malyshev VP (1977) Mathematical planning of metallurgical and chemical experiments. Alma-Ata: Sci 35

Matsuno T, Kadota M, Ishiguro Y (1958) Separation of gypsum by the flotation process. Bull Soc Salt Sci 12(2):73–78

31

Mineralogical and Technological Aspects of Phosphate Ore Processing

A. Elbendari(✉), V. Potemkin, T. Aleksandrova, and N. Nikolaeva

Mineral Processing Department, Saint Petersburg Mining University, Saint Petersburg, Russia
abdullah_elbendary@yahoo.com

Abstract. The article studies the mineralogical features of phosphate ores. In the conditions of declining industrial reserves of apatite-containing ores, issues of a more comprehensive and in-depth study of the mineral and material composition, as well as the improvement of existing technologies for the processing of this type of raw material, become topical. Using optical methods of analysis, electron microscopy with automated mineralogical analysis (MLA), mineral and elemental composition of apatite was obtained. Taking into account the studied mineralogical and material composition, experiments on grinding and flotation were carried out. Based on these data, it was concluded that the optimal scheme for the processing of phosphate ores is a flotation scheme with preliminary selective disintegration.

Keywords: Mineralogical composition · Grinding · Beneficiation · Phosphorus-bearing minerals

1 Introduction

Phosphates are one of the most important minerals on Earth, as they are used as fertilizers for agriculture and as a necessary raw material for the chemical industry (Brylyakov 2004; Abouzeid 2007). In addition, phosphates are the source of rare-earth elements. They are used in many commercial and industrial products, such as: detergents, toothpastes and fireproof materials. Worldwide consumption of P_2O_5 in all of the areas above is projected to grow gradually from 44.5 million tons in 2016 to 48.9 million tons in 2020 (Jasinski 2017).

In the conditions of declining industrial high-quality reserves of phosphorus-containing ores, issues of a more comprehensive and in-depth study of the mineral and material composition, as well as improvement of existing technologies for processing this type of raw material, become urgent. The study of the influence of the mineral raw materials composition on the features of the beneficiation schemes construction is given in the works of many authors (Aleksandrova et al. 2012; Evdokimova et al. 2012; Gerasimova et al. 2018; Litvintsev et al. 2006; Mitrofanova et al. 2017).

2 Methods and Approaches

The object of the study was apatite-nepheline ore of the Khibiny deposits group (Russia) and phosphate ore of the Abu-Tartur deposit (Egypt). For the development of beneficiation schemes and modes, complex studies were carried out on mineral and elemental composition, including optical methods of analysis, electron microscopy using automated mineralogical analysis (MLA), etc. As a result of the work, the mineral composition of apatite-nepheline ore (ANO) and phosphate ore (PO) was studied taking into account the data of optical and electron microscopic studies, spot X-ray spectral and chemical analyzes, atomic emission spectrometry, automated mineralogical analysis. The chemical composition of ANO and PO is given in the Table 1.

Table 1. The chemical composition of ANO and PO

Apatite-nephelineores (Russia)		Phosphoriteore (Egypt)	
Component	Content, %	Component	Content, %
SiO_2	32.50	SiO_2	19.4
TiO_2	2.20	TiO_2	0.06
Al_2O_3	15.52	Al_2O_3	2.6
Fe	5.45	Fe	5.3
CaO	18.28	CaO	37.6
MgO	0.92	MgO	0.8
MnO	0.16	MnO	0.05
K_2O	4.20	K_2O	0.59
Na_2O	5.56	Na_2O	0.57
P_2O_5	12.50	P_2O_5	21.8
SO_3	0.04	SO_3	1.6
CO_2	0.05	CO_2	1.2
LOI	2.62	LOI	8.43
Total	**100.00**	Total	**100.00**

To study the possibility of increasing efficiency of the flotation process, studies were carried out on the selection of the optimal grinding mode and a series of flotation experiments.

3 Results and Discussion

According to the mineralogical analysis, the main primary minerals of ANO are apatite and nepheline, the contents of which are respectively 30.67 and 30.88%; minor quantities contain pyroxenes, mica, feldspars, as well as natrolite and kaolinite – secondary minerals formed due to the destruction of the primary mineral phases.

Phosphorus-containing minerals of the sample are apatite, eschynite, phosphates of rare-earth elements and lomonosovite, with a distribution to these minerals of 99.94, 0.01, 0.02 and 0.03% phosphorus respectively. Valuable minerals of the sample are apatite – the main mineral concentrating phosphorus and nepheline – the main mineral concentrating aluminum.

Apatite is the main mineral concentrator of phosphorus in the ore; it forms disseminated prismatic crystals, vein clusters of crystals, less often massive clusters of fractured xenomorphic grains, often included in the grains of other minerals - pyroxenes, mica, sphene, nepheline (Fig. 1).

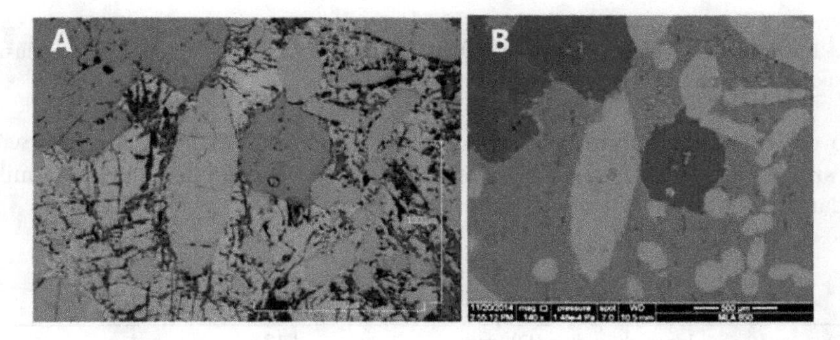

Fig. 1. Apatite crystals in the aegirine-augite matrix. Spectra in (b): 1, 7 - nepheline; 2, 4 - aegirine-augit; 3, 8 - apatite; 5, 6 - arfvedsonit

Image: (a) - in reflected light; (b) - in backscattered electrons. Apatite is characterized by a pronounced idiomorphism of grains that have clear crystallographic outlines; the shape of apatite grains is columnar, prismatic, acicular, which causes a weak connection between them in aggregates. The crystalline form of apatite, the natural brittleness of the mineral will contribute to the primary destruction of the mineral during ore grinding and the concentration of apatite in smaller grades.

According to MLA data, 32.78% of the mineral in ore is distributed into free particles, 22.29% into binary and 44.92% into polymineral intergrowth.

Phosphate minerals in PO are represented by carbonate-fluoroapatite, fluoroapatite, hydroxylapatite, and francolite. The most common clay minerals in the studied sediments are smectites. The amount of kaolinite and illite in general is insignificant, although their content is also high enough. Among the non-phosphate components of the PO, detrital quartz, as well as ankerite and pyrite cement in unaltered phosphates, are predominant. Pyrite and ankerite in many cases replace partially or completely phosphate grains. Elements such as Ba, Cr, Ni, Sr, Y, and Zr are found in relatively high concentrations, while Co, Nb, Pb, Rb, Th, and U are found in relatively low concentrations (Abdel-Moghny and Zhabin 2011; Baioumy 2013) (Fig. 2).

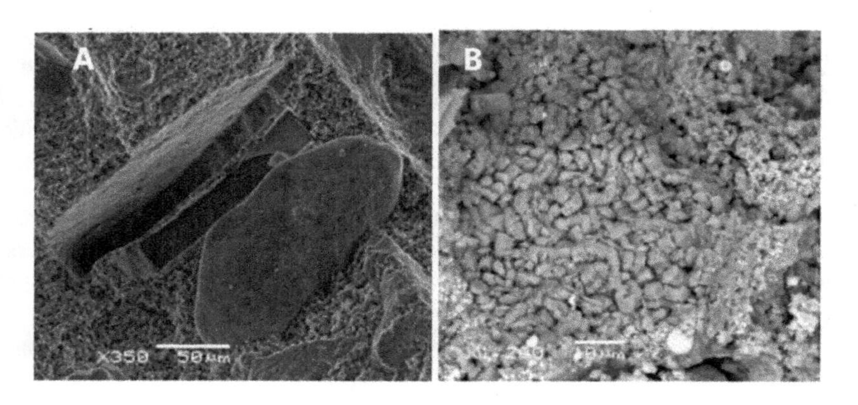

Fig. 2. Electron microscopic images of PO (Baioumy 2013). (a) – rounded silica inclusions; (b) – kaolinite plates

To determine the optimal processing scheme for ANO and PO, samples with a grain size of less than 2 mm and a mass of 550 g were ground in a ball mill in accordance with the specifications given in Table 2:

Table 2. Mill characteristics and experiment conditions

Mill	Inner diameter (D), mm				125			
	Length, mm				170			
	Volume, cm^3				2085			
	Critical speed, rpm				120			
Media (Balls)	Material				Alloy steel			
	d, mm	19	22	25	27	28	29	30
	Number	1	2	2	4	4	3	1
	d, mm	32	34	35	36	39	40	41
	Number	2	1	2	1	1	2	1
	Specific gravity				7.8			
	Mass of balls, g				3388.5			
Material	Igneous & sedimentary phosphate ore							
	Specific gravity				3.1 & 2.7			
	Powder weight, g				550 gm			

The grinding results are shown in Fig. 3.

As a result of the grinding process study, it was found that with an increase in the grinding time, the particle size sharply decreases, and pulp grinding with a solid content of 50% gives the best results. Studies have also been conducted for the process of grinding apatite ore with the addition of tributyl phosphate in amount of 500 and

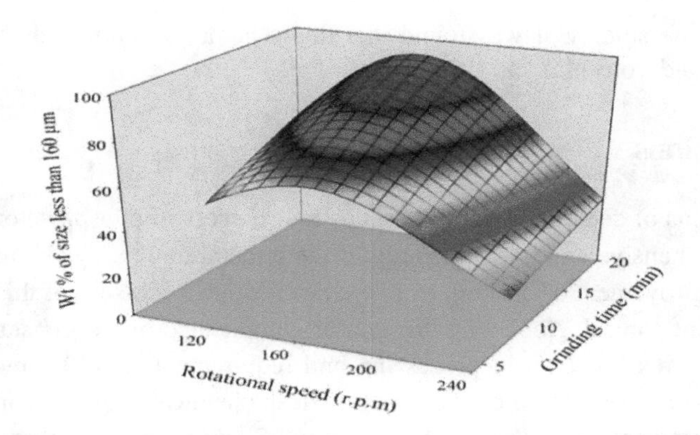

Fig. 3. 3D relationship between the rotational speed and grinding time

1000 ml/ton. Studies have shown that the addition of surface-active substances (tributyl phosphate) during grinding of ANO and PO does not only increase the efficiency of grinding, but also partially convert rare-earth metals into soluble form with their subsequent extraction.

Beneficiation of ANO and PO samples was carried out according to the flowchart shown in Fig. 4.

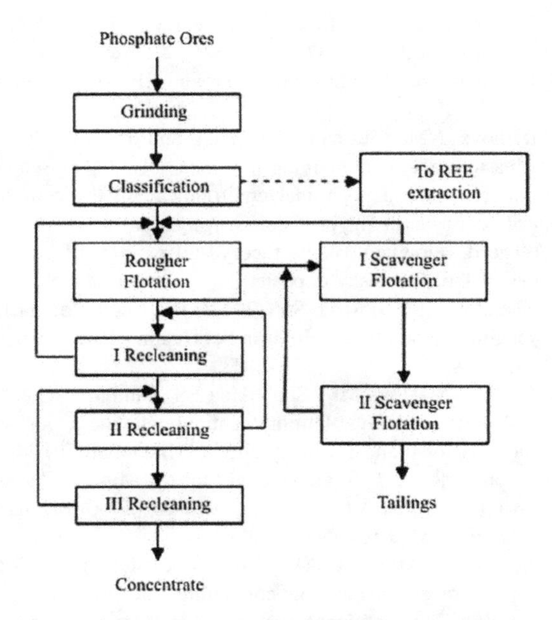

Fig. 4. Block diagram of phosphate ore beneficiation

Based on the studies, it was found that this scheme is optimal for both samples (sedimentary and volcanic).

4 Conclusions

In the conditions of declining quality of industrial reserves of phosphate ores, issues of a more comprehensive and in-depth study of the mineral and material composition, as well as the improvement of existing technologies for the processing of this type of raw material become topical. Achieving this goal is complicated by the constant decline in the quality of ores involved in processing and requirement of 100% recycled water supply implementation. Based on the mineralogical, chemical, material composition, as well as technological research on the possibility of processing phosphate ores, it was concluded that the optimal scheme for the extraction of apatite is a flotation circuit with preliminary selective disintegration. At the same time, the possibility of extracting rare earth metals has been established.

Acknowledgements. The work is carried out under financial support of the Ministry of Education and Science of the Russian Federation, the project RFMEFI57417X0168.

References

Abdel-Moghny MW, Zhabin AV (2011) Mineralogical features of phosphate rocks Egypt. Bull Voronezh State Univ Ser Geol 2011(1):37–49
Abouzeid AM (2007) Upgrading of phosphate ores – a review. Powder Handling Process 19:92–109
Aleksandrova TN, Litvinova NM, Gurman MA, Aleksandrov AV (2012) Comprehensive utilization of the far eastern apatite-containing raw materials. J Min Sci 48(6):1047–1053
Baioumy HM (2013) Effect of the depositional environment on the compositional variations among the phosphorite deposits in Egypt. Geol Geophys 54(4):589–600
Brylyakov YE (2004) The development of the theory and practice of complex enrichment of apatite-nepheline ores of the Khibinsky deposits
Evdokimova GA, Gershenkop ASh, Fokina NV (2012) The impact of bacteria of circulating water on apatite-nepheline ore flotation. J Environ Sci Health - Part A Toxic/Hazardous Subst Environ Eng 47(3):398–404
Gerasimova LG, Nikolaev AI, Maslova MV, Shchukina ES, Samburov GO, Yakovenchuk VN, Ivanyuk GY (2018) Titanite ores of the khibiny apatite-nepheline-deposits: selective mining, processing and application for titanosilicate synthesis. Minerals 8(10):446
Jasinski SM (2017) Phosphate Rock, U.S. Geological Survey, Mineral Commodity Summaries
Litvintsev VS, Melnikova TN, Yatlukova NG, Litvinova NM (2006) Mechanoactivation in the processes of ore preparation. Mt J (6):95–96
Mitrofanova GV, Ivanova VA, Artemiev AV (2017) Use of reagents-flocculants in water-preparation processes during phosphorous-containing ore processing. In: International multidisciplinary scientific GeoConference surveying geology and mining ecology management, SGEM 2017, vol 17, no 11, pp 1143–1150

Topochemical Transformations in Sodium-Bismuth-Silicate System at 100–900°C

A. Pavlenko(✉) and R. Yastrebinskiy

Belgorod State Technological University named after V.G. Shukhov, Belgorod, Russia
yrndo@mail.ru

Abstract. The authors have developed a method for producing highly dispersed sillenite bismuth silicate in the system Na_2O-Bi_2O_3-SiO_2 (NBS) from water solutions of organosilicon monomers (sodium methylsiliconate) and bismuth nitrate. The paper studies the phase composition and microstructure of the synthesized NBS material at different temperatures. The morphology of crystals in the NBS material and the peculiarities of its thermal-oxidative breakdown are investigated. X-ray diffraction spectra obtained using a cuK_α-source are used to evaluate the crystal lattice spacing and to analyze the broadening of the maximum-intensity diffraction line for this crystal with due consideration of crystal indices h, k, l by the approximation method to determine the dimensions of the coherent scattering region and microdistortions of the crystal lattice $\Delta a/a$. The authors established that the silicate shell on $Bi_{12}SiO_{20}$ particles is close to the silicates with continuous chain radicals $[SiO_3]^{2-}_{\infty}$, and a part of them are bridges between the bismuth silicate particles.

Keywords: Sillenite · Phase composition · Microstructure · Crystal morphology · Thermal treatment · Crystal lattice microdeformation

1 Introduction

The development of highly dispersed metal-organosiloxane fillers with modified surface allows solving a multitude of important problems in the field of radiation materials science (Pasechnik 2006). The promising approach is to use water-soluble chemically active organosiloxanes as the basis for production of metal oligomers. A new technological approach to the solution of the stated complex problem is required.

As of today, the chemistry of organosiloxane compounds of bismuth attract particular attention. This is conditioned by multiple valuable properties of organosilicon compounds (high thermal stability, hydrophobicity, dielectric characteristics and resistance to a range of aggressive media). Besides, bismuth atoms have large capture cross-section of gamma-radiation, which is almost the same as for lead atoms in a wide energy spectrum. The presence of vacant 3d-orbitals in silicon atoms conditions high reactivity of bond ≡Si-OH in silicate minerals.

2 Materials and Methods

The authors have developed a method for producing highly dispersed sillenite bismuth silicate in system Na_2O-Bi_2O_3-SiO_2 from water solutions of organosilicon monomers (sodium methylsiliconate) and bismuth nitrate (Yastrebinskii et al. 2018).

The amounts of the components were calculated with the aim of producing stable bismuth silicate $Bi_{12}SiO_{20}$ ($6Bi_2O_3 \cdot SiO_2$).

At 100 °C we have obtained highly dispersed (0.2–0.3 μm) hydrophobic NBS material that is insoluble in water (NBS wetting angle is 122°). The density is 3780 kg/m^3.

According to mass-spectroscopy, NBS material had the following composition (expressed as oxides), wt%: Na_2O - 23.83; Bi_2O_3 - 59.70; SiO_2 - 16.47.

Differential thermal analysis (DTA) and thermogravimetric analysis (TGA) of specimens were performed on a STA-449 F1 Jupiter derivatograph (Germany). X-ray diffraction (XRD) analysis of phases and structure was performed on a ARL™ X'TRA Powder Diffractometer (Switzerland) with Cu_{kx} source (λ_{kx} = 1.542 Å) using a nickel filter. The infrared spectra were obtained on a Specord-75IR spectrometer (Germany). The material microstructure was studied by raster electron microscopy (REM) in the modes of reflected (back-scattered) and secondary electrons.

3 Results and Discussion

Phase Composition and Microstructure of Mineral Phases in NBS Material Synthesized at 100 °C. XRD phase analysis using the Powder Diffraction File (PDF) and literature data (Gorshkov and Timashev 1981; Gorelik et al. 2002) allowed detecting the formation of three amorphous-crystalline mineral phases:

1. Metastable bismuth silicate Bi_2SiO_5 (d = 3.0379 Å/I = 100%; 3.7169 Å; 2.7223 Å) with tetragonal crystal system (a = 3.802; c = 15.134 Å), with the amorphous ring of about 3 Å.
2. Bismuth oxide α-Bi_2O_3 (d = 3.2596 Å/I = 100%; 3.2596 Å; 1.9625 Å) with monoclinic crystal system (a = 5.8499; b = 8.1698; c = 7.5123 Å) with the amorphous ring of about 3 Å.
3. Bismuth organosilicate $H_3C(Si_xBi_yO_z)Na$ with the amorphous ring of 10–12 10–12 Å and clear X-ray reflection at d = 11.4513 and 5.7090 Å. However, the precise determination of this composition using PDF failed.

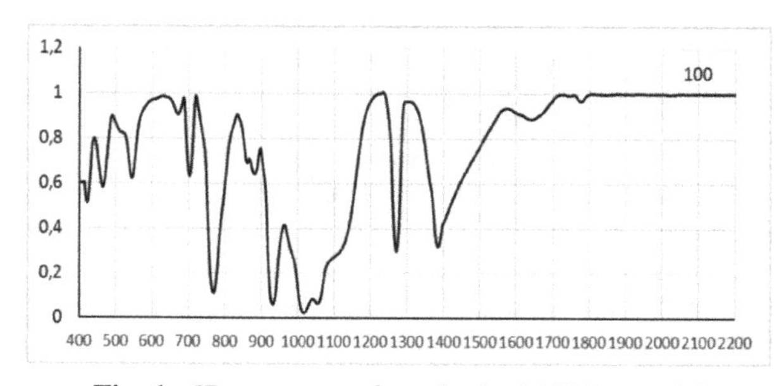

Fig. 1. IR spectrum of synthesized NBS material

The results of IR-spectroscopy, the silicate phases in NBS powder synthesized at 100 °C have linear structure. The splitting of the absorption bands in the range of 1000–1100 cm^{-1} that is typical for the siloxane bond indicates the presence of several types of siloxane phases (Fig. 1).

Morphology of Crystals in Synthesized NBS Material. According to REM, NBS material synthesized at 100 °C contained particle agglomerations of irregular shape with the size of 0.8–2.5 μm (Fig. 2).

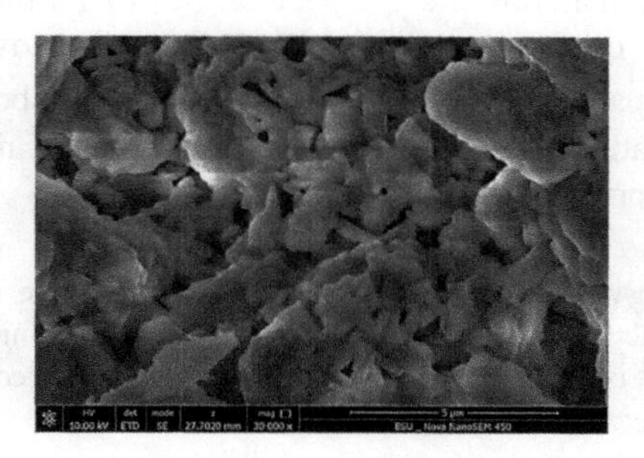

Fig. 2. Electron microphotographs (REM) of NBS material synthesized at 100 °C

Defectiveness of Crystals in NBS Material Subjected to Thermal Treatment. X-ray diffraction spectra obtained using a CuK_{α}-source were used to evaluate the crystal lattice spacing and to analyze the broadening of the maximum-intensity diffraction line for this crystal with due consideration of crystal indices h, k, l by the approximation method to determine the dimensions of the coherent scattering region and microdistortions of the crystal lattice Δa/a.

At 100–300 °C XRD analysis detected amorphous-crystalline bismuth organosiliconate $H_3C(Si_xBi_yO_z)Na$ with the amorphous ring of 10–12 Å and clear X-ray reflection at d = 11.4810 Å and 5.7020 Å.

At the temperature of 200 °C, bismuth silicate Bi_2SiO_5 in the mixture of minerals in terms of X-ray parameters approaches to the benchmark silicate of this composition (card no. 36-288 PDF: d = 3.0400 Å (I = 100%, hkl = 103).

In the temperature interval of 300–400 °C, metastable bismuth silicate $Bi_{12}Si_{0,87}O_{20}$ with cubic crystal lattice in the synthesized dry mix transforms into stable bismuth silicate $Bi_{12}SiO_{20}$ also with cubic lattice.

In the temperature interval of 300–500 °C, the density of dislocations in the structure of bismuth silicate $Bi_{12}SiO_{20}$ crystals was lowering, while at the temperature higher than 650 °C, it was conversely rising up. The increase of the temperature from 300 to 500 °C improves the parameters of elementary crystal lattice of bismuth silicate $Bi_{12}SiO_{20}$ by 0.0357 Å and 0.0268 Å, as compared to the benchmark crystal. The volume of elementary crystal cell in this temperature interval grows by 2% and amounts to 1040.5870 $Å^3$.

4 Conclusion

In the study, a method for producing highly dispersed sillenite bismuth silicate in the system Na_2O-Bi_2O_3-SiO_2 (NBS) from water solutions of organosilicon monomers (sodium methylsiliconate) and bismuth nitrate was developed. The paper studied the phase composition and microstructure of the synthesized NBS material at different temperatures. The crystalline structure of the substance and the presence of silicate amorphous phase in it were discovered. The paper revealed the morphology of crystals in the synthesized NBS material and the peculiarities of its thermal-oxidative break-down; the silicate shell on the particles of $Bi_{12}SiO_{20}$ was close to continuous chain radicals $[SiO_3]_{\infty}^{2-}$ and a part of them played the role of bridges between bismuth silicate particles. The determination of physicochemical properties of modified $Bi_{12}SiO_{20}$ sil-lenite crystals was of appreciable significance.

Acknowledgements. The work is realized in the framework of the Program of flagship university development on the base of Belgorod State Technological University named after V.G. Shukhov, using equipment of High Technology Center at BSTU named after V.G. Shukhov, the project within strategic development program No. a-51/17.

References

Gorelik SA, Skakov YuA, Rastorguev LN (2002) X-radiography and electron-optical analysis (in Russian). MISIS, Moscow, 360 p

Gorshkov VS, Timashev VV (1981) Methods of physicochemical analysis of silicates (in Russian). Higher School, Moscow, 335 p

Pasechnik OF (2006) Study of properties and structure of polyamide films after interaction of low-orbit space factors. Obninsk, p 113. (in Russian)

Yastrebinskii RN, Bondarenko GG, Pavlenko VI (2018) Synthesis of stable bismuth silicate with sillenite structure in the Na2O–Bi2O3–SiO2 system. Inorg Mater Appl Res 9(2):221–226

Comparative Gold Deportment Study on Direct Leaching and Hybrid Process Tails of Oxide Ores from Mayskoye Au Deposit

I. Anisimov(✉), A. Dolotova, A. Sagitova, M. Kharitonova, B. Milman, and I. Agapov

Science and Technology Research Division, Polymetal Engineering, Saint-Petersburg, Russia
anisimovis@polymetal.ru

Abstract. Mayskoye gold deposit is located in Chukotka, Russia. Ore bodies are mineralized brecciation zones composed of vein-quartz, argillic altered rocks (siltstone and carbonaceous silts) with fine disseminated and veinlet gold-bearing arsenopyrite and pyrite. Two main technological types of ores were distinguished at the deposit: primary and oxide. The main reserves were represented by primary ores, which are classified as refractory. Oxide ore had a quartz-micaceous composition with minor feldspars, kaolinite and sulfides. Beside native visible and colloidal gold, other main carriers of gold in oxide ore are arsenopyrite, pyrite, minerals of scorodite group and stibnite. Tested oxide ore sample showed low recoveries, according to the existing flow sheet in the CIL plant. Cyanidation tests showed some preg-robbing effect on organic matter. Flotation of carbonatious matter with consecutive leaching of flotation tails proved to have better recoveries than direct leaching by reducing preg-robbing on carbon three times.

Keywords: Refractory gold · Preg-robbing · Invisible gold · Surface contamination

1 Introduction

Mayskoye Au deposit is located in Chaunsky region in Chukotka, Russia. The ore was formed by quartz-pyrite-arsenopyrite shear opening veins in terrigenous sequence of carbonaceous siltstone and sandy-siltstones. Ore bearing sequence altered to micaceous-carbonate-albite rocks (of beresite formation). Thus, the ore had quartz-micaceous composition with minor feldspars and sulphide material. Major part of gold was refractory and concentrated in arsenopyrite and partially pyrite. Thus primary ores are treated with flotation recovering around 90% of gold with sulfide concentrate, which was shipped to Amursky POX plant for oxidation and cyanide tank leaching.

Part of the carbonaceous matter is preg-robbing gold. Upper part of the deposit contains oxide ore, which mostly was not refractory and free leaching. The carbon in pulp plant treats oxide ore during the warm season.

The studied oxidized ore sample had low gold recovery in CIL plant, thus directed to gold deportment analysis and lab test work. The sample mineral composition was quartz-micaceous with minor kaolinite and secondary minerals scorodite, jarosite, tripuhyite and others. Natural coal content was 1,4%. Sulfides were represented by pyrite, marcasite, arsenopyrite and stibnite. Oxidized ore appeared to be partially refractory. Most of the refractory gold was represented by "invisible" gold. It was contained both in the native form, and was dispersed in liberated ore minerals: pyrite, arsenopyrite, stibnite and scorodite, and their binaries with gangue. The form of occurrence of gold in the ore and its connection with mineral carrier was important to determine effective methods for metal extraction. Such study in plant products was necessary to chase the issue and find optimization in reagent scheme or the flow-sheet.

2 Samples and Methods of Study

Two samples of initial oxidized ores from 1st and 2nd ore zones of Mayskoe gold deposit and four plant products: direct cyanidation tails, flotation tails and combined leach tails of flotation tails leaching product. The examination methodology included the following stages: classification of samples into narrow size classes, separation of material in heavy liquids with density of 2,90 g/cm^3 and 2,5 g/cm^3. Gold concentrate was panned from the heavy concentrate. The following separation products were obtained: gold gravity concentrate, heavy concentrate, rock-forming lights and carbonaceous fraction. Bulk mineral composition of the initial size classes and lights was studied by powder X-ray phase analysis using the Rietveld refinement. The mineral composition of the gold concentrates and heavy concentrates were studied by optical microscopy and the mineral identification of the carbonaceous fraction was done with stereo microscope.

Gold content in the products was analyzed by fire assays; arsenic, antimony and other elements by XRF in all products, except for gold concentrates. In the gold tips, the quantitative finding of gold was carried out by optical and electron microscopy.

Chemical composition of minerals-carrier of gold and other ore minerals was studied with SEM-EDX.

3 Results and Discussion

Quartz and phyllosilicates (muscovite, illite, smectite, kaolinite, dickite) were the main minerals in the samples; scorodite, jarosite, sulfides and other minerals were accessories. The main minerals-carrier of gold are represented by arsenopyrite, pyrite, marcasite and stibnite, oxidized forms of arsenic - scorodite, and antimony - tripuhyite, stibiconite, cervantite and valentinite. Native carbon content in the ore was 0,1–1,4%.

Visible native gold in samples had high fineness (902–914‰). The distribution pattern of visible native gold particles in ore zone 1 and 2 were slightly different. 70% of the visible native gold from head sample from the first ore zone was larger than 90 μm. 65% of the visible native gold from the second ore zone head sample was coarser than 70 μm. All visible gold particles in the cyanidation tails of first ore zone were less than 45 μm. Gold particles coarser than 45 μm dominated in cyanidation tails of the sample from the second ore zone. Native gold observed in the samples of direct leach tails were found in free particles with the surface shielded or in binaries with oxidized antimony forms. Rarely binaries of gold with pyrite, stibnite and gangue minerals were observed. Fineness of visible gold in cyanidation tails was 10–15% lower than in initial products with higher silver content. Besides natural gold the metal was in dispersed form in the mineral-carriers: pyrite, arsenopyrite, stibnite, scorodite, antimony oxidized forms and bound with carbon (Fig. 1).

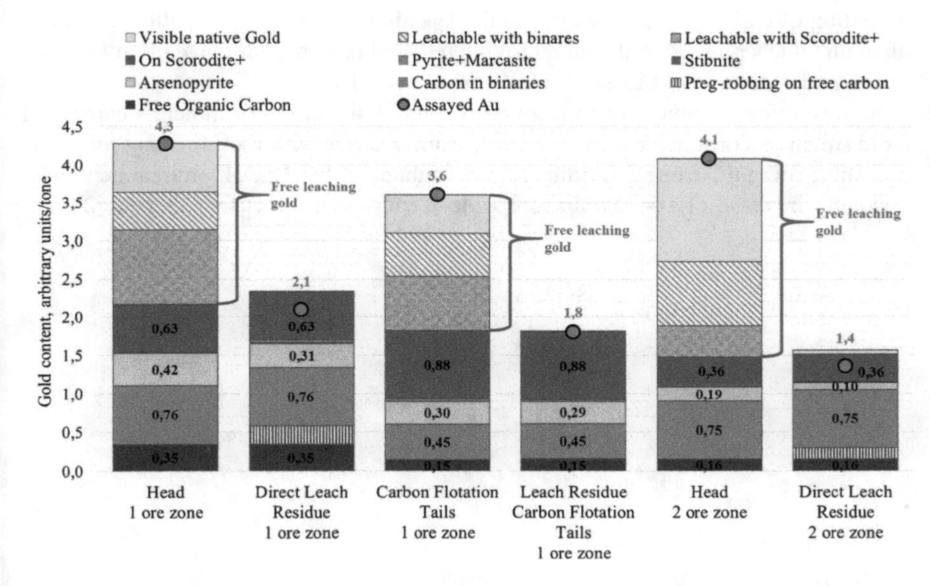

Fig. 1. Gold deportment in the plant feed, direct leach tailings, flotation tails and flotation tails leach residues of oxidized ores from 1st and 2nd ore zones of Mayskoe gold deposit

Free leaching gold accounted for 49% in head sample from the first ore zone and 64% in head sample from the second ore zone. The rest of the gold was on carbon (about 22–26%), in sulfides (about 5–10%) and scorodite (about 9–15%). Visible gold extraction by direct cyanidation was in range of 96–99% and gold from scorodite only 53–61%. The losses of visible gold in leaching occurred due to the blocking of the surface with compounds of oxidized antimony, as well as the slower cyanidation of gold with lower fineness. Coal flotation reduced gold losses by 22%.

4 Conclusions

Gold in the head samples was represented by both visible high-grade gold, most of which was a of gravity size and invisible/colloidal gold (<0,5 micron), forming more than half of the metal of the original ore. Lower grade gold particles leached slower.

The oxide ore of zones 1 and 2 of Mayskoye deposit was partially refractory, since most of the gold was invisible and dispersed in arsenopyrite, scorodite and with carbon.

The main gold losses during cyanidation were associated with invisible gold enclosed in sulfides, scorodite and organic carbon.

Based on gold deportment results obtained, the following suggestions were made:

1. Introduction of gravity in the grinding cycle would help reduce the loss of coarse gold and extract some of the refractory gold with large particles of arsenopyrite and stibnite. Increase in recovery can vary as coarse gold content and be as much as 2–5%.
2. Addition of lead nitrate in cyanidation process may increase the dissolution of low-grade gold particles.
3. Flotation of carbon with consecutive tails leaching proved to have better recoveries than direct leaching. Partial coal removing reduced preg-robbing on carbon for three times and reduced gold losses by 22% from zone 1 ore.
4. The most effective objective to recovery would be liberation of minerals-carriers of gold (mainly scorodite and arsenopyrite) from binaries with rock forming minerals.
5. Leaching of gold from scorodite can be enhanced by NaOH pretreatment with recovery increase of 10–15% for ore zone 1 and lesser effect for ore zone 2.

34

Applied Mineralogy of Anthropogenic Accessory Minerals

A. Gerasimov[1(✉)], E. Kotova[2], and I. Ustinov[1]

[1] REC «Mekhanobr-Tekhnika», St-Petersburg, Russia
gornyi@mtspb.com

[2] Mining Museum, St-Petersburg Mining University, St-Petersburg, Russia

Abstract. The term "accessory minerals" is applicable to an expanding range of mineral resources, including rare minerals of anthropogenic deposits and mineral metallurgical wastes. Typical accessory minerals of sedimentary rocks, metallurgical slags and calcines include sulphide minerals of iron, copper, and zinc. This work covers certain chemical transformations of accessory sulphides contained in the wastes of the metallurgical industry. Note that similar behaviors of sulphides may also be observed in the thermal processing of coal. It is shown that, in terms of thermodynamics, thermal processing of products containing precious metals enables application of various methods for generating artificial covellite with silver and gold content through the use of accessory sulphide minerals.

Keywords: Accessory minerals · Precious metals · Thermal processing · Thermodynamics

1 Introduction

Accessory minerals, in the classical sense, are minerals contained in rocks in small quantities (less than 1%). It may be assumed that the term is also applicable to an expanding range of mineral resources, including rare minerals of anthropogenic deposits and mineral metallurgical wastes. Typical accessory minerals of sedimentary rocks, metallurgical slags and calcines include sulphide minerals of iron, copper, and zinc, often containing precious metals.

This work covers certain chemical transformations of accessory sulphides contained in the wastes of the metallurgical industry. Note that similar behaviors of sulphides may also be observed in the thermal processing of coal. Process mineralogy (Bradsow 2014), recently expanded to include thermal processes (Rubin et al. 2014), is the critical tool when studying the behavior of accessory sulfide minerals.

2 Methods and Approaches

The traditional technology used for the processing of zinc concentrates obtained in flotation concentration of polymetallic sulphide ores envisages oxidizing roasting of the zinc concentrate, in which the main mineral, ZnS sphalerite, is converted into zinc sulphate $ZnSO_4$. The inevitable impurities of pyrite FeS_2 and chalcopyrite $CuFeS_2$, present in the zinc concentrates, are also oxidized into the corresponding simple sulfates and oxides at temperatures above 770 °K (Naboichenko et al. 1997). At present, fluidized bed furnaces are predominantly used for oxidizing roasting, with the oxidation process occurring in a hot gas flow in a suspended state.

Despite the obvious benefits, such as high yield and roasting rates, fluidized bed furnaces also have certain disadvantages, including partial underburning of sphalerite due to the rapid formation of a poorly gas-permeable zinc sulfate film on its surface. This inevitably leads to underburning of fine inclusions in sphalerite grains of other sulfide minerals, primarily chalcopyrite. Chalcopyrite is the most common impurity in sphalerite, which is due to the high similarity of the structures of their crystal lattices. It is known that chalcopyrite is most frequently found in the sphalerite structure in the form of very fine grains, down to emulsion shots (Betekhtin 2008). The behavior of chalcopyrite inclusions in sphalerite is significant for subsequent stages of processing of the calcined concentrate.

After the oxidizing roasting, the resulting zinc calcine is subjected to staged sulfuric acid and aqueous leaching of zinc sulfate and oxide-sulfate with the recovery of a zinc sulfate solution to be subsequently used for zinc metal precipitation by electrolysis. Acid and neutral zinc sulfate leaching is performed at temperatures of up to 80–100 °C, additionally enabling chemical transformations of insoluble calcine components on the grain surfaces. The insoluble cake obtained after leaching includes a certain amount of underburned sulphide minerals containing precious metals, primarily silver. It was previously demonstrated (Otrozhdennova et al. 1997; Geikhman et al. 2003) that sulphide minerals remaining in the cake may be flotation concentrated, yielding a froth product containing up to 18% of sulfide sulfur and up to several kilograms of silver per ton.

We have studied the composition of concentrates obtained by flotation concentration of the cake generated after zinc sulphate leaching at three metallurgical enterprises in different countries. The studies were performed using a Camscan scanning microscope, an Analysette laser microanalyzer, a Geigerflex X-ray phase spectrometer, optical microscopy tools, etc.

It was found that there was no fundamental difference in the material composition of such concentrates at different metallurgical plants. This is explained by the similarity of the ore bases and of the metallurgical treatment structures used at the mining enterprises analyzed. Sphalerite (70–80%) is the main mineral found in the composition of the concentrates, followed by such spinel-structure minerals as $MeFe_2O_4$, covellite, oxide compounds of zinc and iron (several percent), and traces of chalcosite.

In terms of their elementary chemical composition, the concentrates were represented by zinc 45–48%, total sulfur 20–22%, sulfur sulfide 17–19%, iron 11–14%, silica 2.5–4%, and copper 3–4%. The silver content was 0.2–0.4% (2–4 kg/t); the gold content was 10–15 g/t.

3 Results and Discussion

The most important finding is the presence of copper sulphide covellite CuS in the flotation concentrate, in which the bulk of silver is concentrated. The availability of silver-containing films of copper sulfides (without their identification) on the surfaces of sphalerite contained in the zinc cake flotation concentrate has been indicated previously (Otrozhdennova et al. 1997). We have established that covellite forms distinct substances on the sphalerite surface. No native silver was found in the samples. Gold is present in the form of fine grains containing silver and copper. The gold sample is 700–930. Copper-containing metals in the source zinc concentrate entering the roasting process are represented almost exclusively by chalcopyrite, including its aggregates with sphalerite.

Two main methods of covellite formation in the metallurgical processing of zinc may be considered in this regard. The first is the interaction of dissolved copper sulfate with the sphalerite surface in sulfuric acid leaching of zinc calcines, with the formation of covellite films in the course of the resulting exchange reaction and with the transition of zinc ions into the solution. The second possible source of covellite is the transformation of fine chalcopyrite inclusions contained in sphalerite grains with the migration of the resulting covellite to the sphalerite surface due to the strong differences in the structures of their crystal lattices. None of these methods has been subjected to special experimental verification. Given that even isomorphous substitution of copper with silver can occur in chalcopyrite, despite the difference in the electron spiral radii of silver of 1.26 Å and copper of 0.96 Å, it may be assumed that silver-containing covellite may be generated using these methods, especially considering the effects produced on such processes by associated minerals.

Let us consider the thermodynamic feasibility of the two hypothetical processes of formation of man-made covellite. In the below calculations of the free energy of the processes and their enthalpy and entropy increments, known reference data are used for reactions (I) and (II). These calculations show that reaction (I) can occur spontaneously, which is facilitated by the entropy effect, and reaction (II) requires external energy inputs, which does not contradict the conditions of respective production processes (Tables 1 and 2).

Table 1. Thermodynamic characteristics

Substance	ΔH^0, kJ/mol	S^0_{298}, J/mol K	C_P, kJ/mol K
ZnS	−206	57.7	46.0
CuSO4	−771.4	109.2	98.9
CuS	53.1	66.5	47.8
ZnSO4	−982.8	110.5	99.2
Реакция	−58.5	10.1	2.1

Table 2. Thermodynamic characteristics

Substance	ΔH^0, kJ/mol	S^0_{298}, kJ/mol K	C_P, kJ/mol K
$CuFeS_2$	−194.93	124.9	95.67
CuS	53.10	66.5	47.8
FeS	−100.00	60.3	50.5
Реакция	41.8	1.9	2.63

$$ZnS + CuSO_4 \rightarrow CuS + ZnSO_4 \tag{1}$$

$$\Delta H^0_{798} = \Delta H^0_{298} + \int_{298}^{798} \Delta c_p dT = -58.5 \cdot 10^3 + 2.1 \cdot (798 - 298) = -57.45\,\text{kJ/mol}$$

$$\Delta S^0_{798} = \Delta S^0_{298} + \int_{298}^{798} \frac{\Delta c_p}{T} = 10.1 + 2.1 \cdot \ln\frac{798}{298} = 12.17\,\text{J/molK}$$

$$\Delta G^0_{798} = \Delta H^0_{798} - 798 \cdot \Delta S^0_{798} = -57450 - 798 \cdot 12.17 = -67.2\,\text{kJ/mol}$$

$$CuFeS_2 \rightarrow CuS + FeS2 \tag{2}$$

$$\Delta H^0_{798} = \Delta H^0_{298} + \int_{298}^{798} \Delta c_p dT = 43,15\,\text{kJ/mol}$$

$$\Delta S^0_{798} = \Delta S^0_{298} + \int_{298}^{798} \frac{\Delta c_p}{T} = 4,4\,\text{J/molK}$$

$$\Delta G^0_{798} = \Delta H^0_{798} - 798 \cdot \Delta S^0_{798} = 39,56\,\text{kJ/mol}$$

4 Conclusions

In terms of thermodynamics, combined thermal processing of products containing precious metals enables application of various methods for generating artificial covellite with the participation of accessory sulphide minerals. Respective statements may be useful for both metallurgical processes and thermochemical processing of hard coals, which always contain impurities of sulphide minerals.

Acknowledgements. The work was carried out with the support of the Russian Science Foundation (project No. 17-17-30015).

References

Betekhtin AG (2008) Course of Mineralogy. Publishing House University, Moscow (2008)

Bradsow D (2014) The role of 'Process mineralogy' in improving the process performance of complex sulfide ores. In: Proceedings XXVII IMPC, C.14, pp 1–23

Geikhman VV, Kazanbaev LA, Kozlov PA (2003) Industrial tests of flotation of zinc cakes. Nonferrous Met (1):29–32

Otrozhdennova LA, Maksimov II, Khodov NV (1997) Combined technology of silver recovery from zinc cakes. Min J (4):39–40

Naboichenko SS (1997) Processes and devices of non-ferrous metallurgy. USMU, Ekaterinburg

Rubin S, Aksenov A, Senchenko A, Lagutina S (2014) Argentojaroside synthesis and research. In: Proceedings of the XXVII IMPC, C.19, pp 209–213

35

Mineralogical Reasons of Au Recovery Variability from North-Western Pit of Varvara Au-Cu Mine (Kazakhstan) and Criteria for Geometallurgical Mapping

I. Anisimov(✉), A. Dolotova, A. Sagitova, M. Kharitonova, and I. Agapov

Polymetal Engineering JSC, Saint-Petersburg, Russia
anisimovis@polymetal.ru

Abstract. Varvara Au-Cu mine deposit is located in Northern Kazakhstan. Mineralization is hosted in volcano-sedimentary, sedimentary rocks, metamorphosed and altered ultramafic and felsic rocks. Variability study was done on 58 composite samples represented five mineral ore types: serpentine-chlorite-talc; carbonate-chlorite-talc; quartz-sulfide; pyroxene-chlorite-prehnite ± garnet; quartz-feldspar ± pyroxene ± amphibole. Five processing ore types were defined: Au, Ni-As, pyrite, Cu and mixed. Mineralogy and geochemical studies revealed separate mineral associations carrying Cu, Ni-As and Au mineralization. Flotation and cyanidation tests were performed for each sample. Au losses with cyanidation cake occurred due to locking in sulfides. Floatation concentrate contamination with Mg-silicates (talc, serpentine) was connected to Au losses. Quartz-sulfide ore demonstrated better recovery by flotation. Cyanidation were most effective for pyroxene-chlorite-prehnite ± garnet and quartz-feldspar ± pyroxene ore compositions. Carbonate presence in the serpentine-chlorite-talc ore followed decrease in recovery by both extraction methods. Optimal viable ore treatment method can be chosen based on regression equations using ore chemical composition and color.

Keywords: Typification · Cyanidation · Flotation · Recovery estimation · Geometallurgy

1 Introduction

Varvarinskoe Au-Cu deposit is located in Kostanay region in Northern Kazakhstan and operated by several open pits. It is localized in volcanogenic, terrigenic, carbonate rocks (D_2–C_1) with ultrabasic and granodiorite intrusions (C_2–C_3). Silicification, argillic alteration, scarn processes were wide spread. Retrograde metamorphic changes were typical for ultramafic rocks.

This work was aimed at determining compositional differences in ore types and ore characteristics effected gold recovery by cyanidation and flotation.

2 Methods and Approaches

58 small composite geometallurgical samples from Northwestern open pit were studied for bulk chemical and mineral composition, sample color with RGB-parameters. Cyanidation and sulfide flotation tests were done on the head samples.

Au was assayed with fire assay with atomic absorption finish, multi-element ICP-AES assays after four acid digestion of samples and XRF-analysis, sulfide and total S, total C estimated by LECO AES analysis.

Mineral phase identification and their quantification was done using Eva software and COD database. Quantitative X-ray powder diffraction with Rietveld refinement was done using Topas software at Polymetal Engineering.

Multivariate statistical analysis was performed on filtered data with Aitchison transformation using Pearson correlations with Cytoscape software, PCA and regression analyses. Regression analysis was carried out to predict gold extraction.

3 Results and Discussion

Mineral composition of the samples varied between serpentine-talc-chlorite ± carbonate, quartz-feldspatic and quartz-sulphide associations. Five main mineral ore types were distinguished based on bulk mineral composition: 1 – serpentine-chlorite-talc, 2 – carbonate-chlorite-talc, 3 – quartz-sulfide, 4 – pyroxene-chlorite-prehnite-garnet, 5 – quartz-feldspar-pyroxene-amphibole.

Flotation and leaching tests revealed five processing ore types: a - gold, b - copper, c - Ni-As, d - pyrite and e - mixed.

Multivariate statistics analyses were performed on all available samples parameters including mineral and chemical composition; color (RGB, brightness – lBRT and darkness - dBRT); material fineness (−71 μA); Au, Cu, As, Ni, sulfidic S, total S recovery to concentrate and losses to flotation tails (XF) (εAuKF, εCuKF, εAsKF, εNiKF, εSsKF, εSKF; εAuXF, εCuXF, εAsXF, εNiXF, εSsXF, εSXF) and cyanidation (εAuCN, εCuCN, −εAuCN, −εCuCN), parameters of cyanidation conditions (reagents uptake - NaCNi, CaCNi), cyano-soluble copper – αCuCNr, Au content in cake – AuXvA, characteristics of gravity con (mass pull - γgk-t, pyrite, arsenopyrite, Cr-spinel, chalcopyrite, Ni-minerals contents in gravity con: PyMrcGK, ApyGK, ChrGK, CpyGK, NiMGK).

Pearson correlations revealed occurrence of 3 geochemical and mineral clusters following bulk sample mineralogy (Fig. 1): quartz-sulfide, diorite (feldspatic) and serpentinite. Au losses with cyanidation tails connected with Au locked in sulfides, which was proved with SEM study. S and Cu contents and NaCN consumption were included in the cluster with Au losses with cyanidation. This fact may point to consumption of free CN^- by reaction with S^0 and Cu dissolution, thus leading to deficit of free CN^-. This reaction produced CuCN and rhodanates that could block surface gold particle impeding its dissolution. Concentrate contamination with self-floating Mg-silicates (talc, serpentine) was connected to Au losses with flotation tails.

Principal component analysis exposed 7 principal components, which explained 72.1% of the total variance and described mineral composition, geochemical

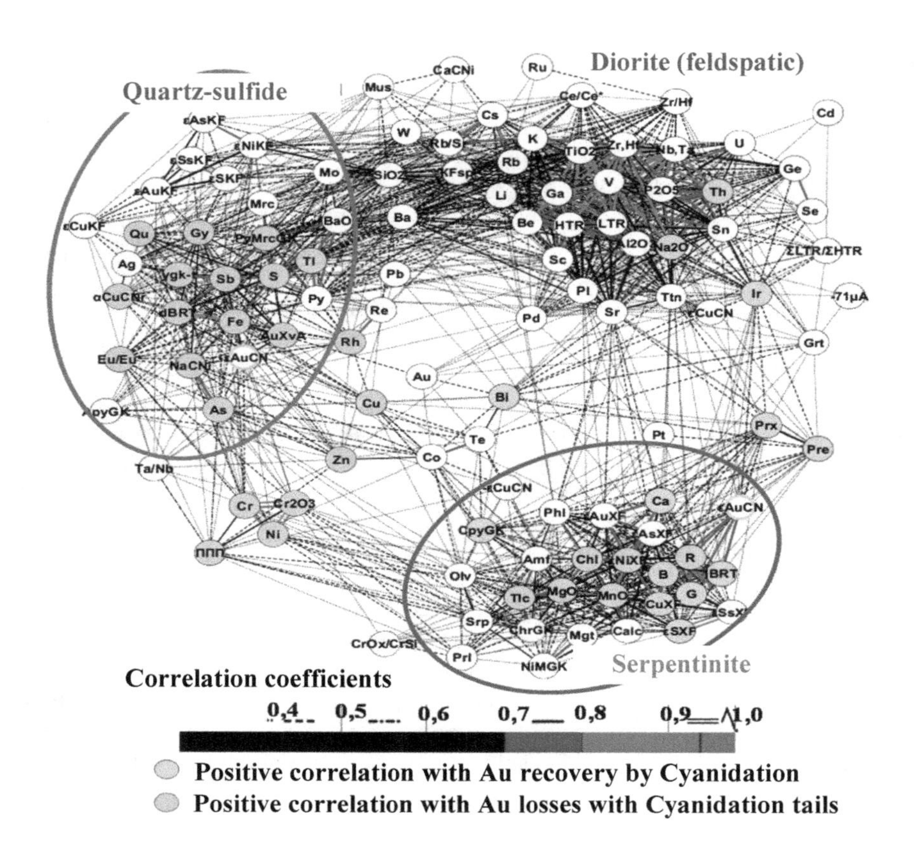

Fig. 1. Positive correlations between compositional and metallurgical parameters of samples from Varvara

associations, Au and Cu recovery by flotation and cyanidation (Fig. 2). 5 ore types were well separated in the coordinates of factors 2 and 3, 2 and 4. Quartz-sulfide ore samples demonstrated better recovery by flotation, while cyanidation was more effective for samples with pyroxene-chlorite-prehnite-garnet and quartz-feldspar-pyroxene-amphibole compositions. Carbonate occurrence in the serpentine-chlorite-talc ore reduced recovery by both concentrating methods.

Regression equations for Au recovery by flotation ($R^2 = 0.80$) and by cyanidation ($R^2 = 0.88$) included color parameter (BRT) and elements contents (ppm or wt%):

$$\boldsymbol{\varepsilon}\mathbf{AuFl} = 127.32 + 0.17\mathrm{Co} - 0.01\mathrm{Cu} - 4.78\mathrm{Fe}(\%) + 5.19\mathrm{K}(\%) + 0.39\mathrm{Mo} + 0.39\mathrm{Th} + 0.01\mathrm{Sb} - 0.63\mathrm{BRT} + 0.10\mathrm{Zn} + 0.43\mathrm{Mg}(\%) + 0.84\mathrm{SiO2}(\%) - 0.81\mathrm{Li} - 0.40\mathrm{Rb} - 1.97\mathrm{Te} + 5.03\mathrm{Cs}$$

$$\boldsymbol{\varepsilon}\mathbf{AuCN} = 226.60 - 2.86\mathrm{Ca}(\%) + 0.30\mathrm{Co} - 7.00\mathrm{Fe}(\%) - 2.08\mathrm{Sc} - 0.16\mathrm{Sr} + 0.13\mathrm{V} - 0.24\mathrm{MgO}(\%) - 3.37\mathrm{Al2O3}(\%) - 0.55\mathrm{SiO2}(\%) + 1.98\mathrm{Ga} - 0.42\mathrm{BRT} + 0.27LOI(\%) + 0.004Cu + 0.02Ni - 0.01As$$

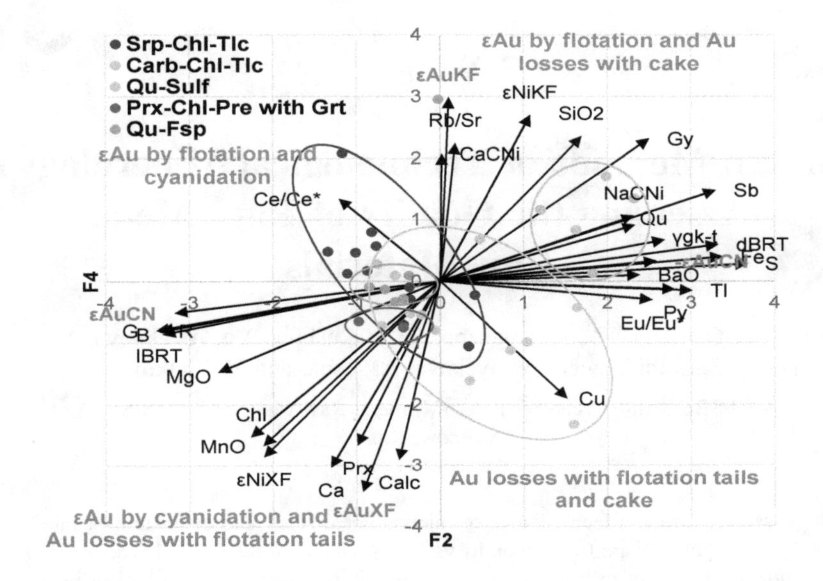

Fig. 2. Factor loadings and factor scores for factors 2 and 4 with interpretation

4 Conclusions

Mineral composition of the samples varied significantly from serpentine-talc-chlorite ± carbonate to quartz-feldspar, skarn and quartz-sulfide associations. Au losses with cyanidation tails connected with Au locked in sulfides. Au losses with flotation tails connected to concentrate dilution with self-floating Mg-silicates (talc and serpentine). Ore sorting can be done based on regression equations of element composition and ore color to direct the ore to more economically viable process.

Acknowledgements. The authors are grateful to Nicolay Rylov for scientific advices, Lims-lab and Nati-R for collaboration.

36

Modern Methods of Technological Mineralogy in Assessing the Quality of Rare Metal Raw Materials

E. Ozhogina(✉), A. Rogozhin, O. Yakushina, Yu. Astakhova, E. Likhnikevich, N. Sycheva, A. Iospa, and V. Zhukova

FSBE "All-Russian Institute of Mineral Raw Materials" (VIMS), Moscow, Russia
vims@df.ru

Abstract. Modern technologies of mineralogical study and evaluation of rare metal raw quality are focused on its variety. Methods of the mineral processing, allowing to optimize monitoring of ore properties defining technological processes and quality of expected products are presented. Some examples of rare metal ores mineralogical study are given. The main challenging tasks in rare metal ores quality evaluation are considered.

Keywords: Technological mineralogy · Methods of analysis · Testing · Rare metals · Minerals · Ores · Specific features · Quality assurance · Quality assessment

1 Introduction

Variety of rare metal raw materials is determined, firstly, by a significant amount of industrial minerals, among which there are more than 20 main and about 30 minor and accessory minerals of different genesis; secondly, by the diversity of their genesis: magmatic, pegmatite, greisen, scarn, metamorphic, hydrothermal, sedimentary, hypergeneous. The rare metal-bearing minerals are the ones that contain Tantalum, Niobium, Bismuth, Tellurium, Zirconium, Hafnium, Yttrium, Scandium, Lanthanides, Lithium, Beryllium, Cesium, Rubidium, Strontium, Barium. A lot of rare metal ores, mainly tantalum-niobium, rare-earth phosphate, carbonate and silicate, and frequently zirconium, are radioactive. The vast majority of rare-metal ores are complex, and industrial minerals can be both main, and secondary, subordinate ones. General features of the rare metal-bearing ores are as follows:

- complex texture-structural pattern (a significant number of fine, metacolloidal spots formed by minerals and aggregates of micro-and nanometer size);
- polymineral composition associated with the simultaneous presence of minerals developed in different paragenetic associations;
- variations in chemical composition of the ore-forming minerals caused by chemical elements isomorphic substitutions in their crystal structure;

- mineral grains phase heterogeneity of various origin, namely decay of solid solutions, syngeneic inclusions, zonal growth, multiple stages of generation, partial recrystallization, secondary solid-phase transformations, etc.;
- ore minerals with radioactive elements can undergo transformations resulting in metamictic forms origin (disrupted crystal structure due to radiation damage) or partially metamictic (damaged crystal structure due to radiation).

2 Methods and Approaches

The variety and complexity of rare metal ores mineralogical features identify the necessity to use a set of modern physical research methods of analysis to get reliable data on their composition and structure. This complex of mineralogical methods is individually selected depending on ores features and research tasks.

Mineralogical study of rare metal ores is necessary at all stages of deposits exploration and development. We particularly note the importance of ore mineralogical and analytical study of at the early stages of geological exploration, that allow to carry out technological assessment of raw materials with minimal investment, and at deposit exploration, involving geological and technological mapping for the detailed study of the mineralization zoning, minerals and mineral associations distribution, variations in the ore-bearing phases properties and characteristics, identification of the technological types and species whiting the geological margins of the deposit. The research practice proves great contribution of mineralogical study in deposits investigation and quality assessment.

When mineralogical investigation is the result of a set of implemented methods, including not only usually traditional ones (optical microscopy, radiography), but also precise analysis (analytical electron microscopy, microprobe). Mineral and technological mapping challenge today the Zashikhinsky, Tomtor, Chuktukon and other deposits of rare earth elements.

3 Results and Discussion

The characteristic feature of technological mineralogy is integration/conjunction of research methods and modern technical means/units. It is especially important when studding the rare-metal mineralization, because it is not always possible to uniquely identify industrially valuable minerals, to establish the mineral form of useful components, to identify and study characteristics of minerals appearing in the fine aggregates.

Pyrochlore-Mmonazite-Crandallite ores of the Tomtor deposit differ in specific composition, are rich in content and reserves of REE, Niobium, Yttrium, Scandi- um, Phosphorus and are a non-standard type of rare-metal raw materials. Ores features are the variable granular composition, often high dispersion, polymineral composition, various forms of occurrence the rare earths and niobium-bearing minerals, the character of their localization in close association with Alumiium- Silicate minerals. The ores are formed by polymineral aggregates with variable content of Crandallite and Kaolinite of fine-grained structure. The aggregate cement contains also Siderite, Ilmenorutil,

Anatase, Pyrochlore, iron hydroxides, and other minerals. Most often the aggregates dispose earthy-type structure.

Pyrochlore is the main Niobium mineral, that form both individual octahedral crystals, fragments of rounded and angular forms, as well as aggregates of tiny grains in size ranging from 1 μm to 0.5 mm, in varying degrees transformed by hypergene processes. The rock is distributed in the form of microinclusions, forming a "rash" in the cementing mass of aluminophosphates and silicates. According to microprobe data (X-ray microspectral microanalysis), the main feature of the hypergenic alteration of Pyrochlore is the replacement of Ca and Na by Sr, Ba and Pb; the altered Pyrochlore varieties significantly dominate in the ores. Hypergeneous transformation of Pyrochlore was accompanied by textural transformations, typically clearly manifested in the disintegration of its large crystals into small blocks (Fig. 1). Often the cracks between the separate Pyrochlore individuals are filled by Crandallite group minerals and Apatite, rarely by sulfides.

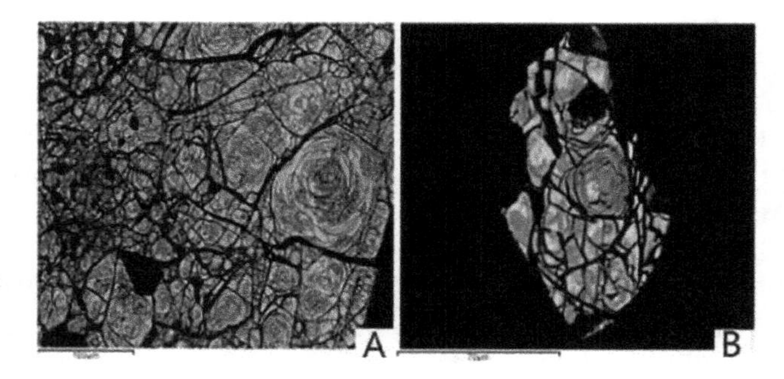

Fig. 1. Disintegrated crystals of Pyrochlore. TEM, image in back scattered electrons

Minerals of Crandallite group mostly form polymineral aggregates being dominant phase in these aggregates. There is a weak individualization of isometric and round shape grains, closely associated with fine Pyrochlore, Monazite, iron hydroxides, their grain size in the ores is often much less than 5 micrometers. According to microprobe analysis, the Crandallite group minerals have a mixed composition. According to X-ray powder diffraction analysis (XPD), the interplanar distances *d*(*hkl*) reliably identificate Gorceixite (2.96, 3.55, 5.72 Å) and the intermediate Goyazite-Florencite series (2, 94, 3, 53, of 5.71 Å).

Mineralogical features of a Pyrochlore-Monazite-Crandallite ores (variable granular composition, often with high dispersion; polymineral composition due to the simultaneous presence of different paragenesis assemblages, different form of Niobium and rare earths presence, vide range of isomorphic substitutions in the structure of minerals, the proximity of their physical properties) determine the impossibility of these ores processing by methods of deep enrichment. Therefore, the prospects of such ores processing should be associated with hydrometallurgy.

Typical features of the Chuktucon Deposit rare earth ores were established on electron microscopic study. The main ore minerals are Pyrochlore, Monazite and the

Crandallite group minerals. All of them are superfinely dispersed and are in close assemblage/association with Iron and Manganese oxides and hydroxides. The latter form complex types of accretions with Pyrochlore (corrosive), Monazite (envelope of Goethite around Monazite grains), Crandallite group minerals (thin jointing), which negatively affects their disclosure and does not allow to identify and study these minerals by traditional methods of optic microscopy. Microprobe study indicated varieties of secondary Pyrochlore (bearing Cerium, Barium, Strontium and mixed type), and variable chemical composition of the Crandallite group minerals. Almost constant presence of Iron and Manganese mechanical impurities was established in all ore minerals. An independent mineral form of Cerium – Cerianite has been identificate by X-ray powder diffraction data.

4 Conclusions

The main challenging tasks for the rare metal ores investigation and quality evaluation should be considered:

- predictive mineral and technological assessment of raw materials of natural and man-made origin;
- geological and technological mapping using an Arsenal of methods of technological Mineralogy;
- forecasting of technological properties of ores at various stages of processing and development of methods of their directed change;
- increase of complexity of development of deposits and deep processing of ores;
- identification and involvement in the industrial use of non-traditional types of rare metal raw materials;
- assessment of environmental consequences of industrial development of deposits.

Therefore, the main task of technological mineralogy in the study of the rare metal ores is today their comprehensive study for quality assessment at all stages of geological research and development of mineral deposits.

37

Mineralogical Breakthrough into Nanoworld

A. Askhabov(✉)

Institute of Geology Komi SC UB RAS, Syktyvkar, Russia
askhabov@geo.komisc.ru

Abstract. A bit more than a hundred years ago Wolfgang Ostwald, Professor at the University of Leipzig, published a book titled "The World of Neglected Dimensions" (Ostwald 1923). He announced a program of a research breakthrough into the world of microscopic particles, which was imminent by that time. A new stage of the intervention into "world of neglected dimensions" began near the end of the 20th century. The main objects at this stage were nanosized particles. The agenda raised issue of development of new sciences, including nanomineralogy. The subsequent mineralogical intervention into the "world of neglected dimensions" proved to be quite successful. We associate challenges of the next breakthrough with the study of objects and processes in the range from individual atoms and molecules to the first mineral individuals (nanoindividuals). The protomineral world is today a new "world of neglected dimensions".

Keywords: "world of neglected dimensions" · Nanoparticles · Quatarone concept · Protomineral world

1 Introduction

The unprecedented interest in nanoscale objects, which we witness in recent years, I called a new stage of intervention into the "world of neglected dimensions". The first stage of the intervention (research breakthrough) occurred at the beginning of the 20th century and was associated with W. Ostwald, who actually referred to the region of microscopic particles as "the world of neglected dimensions" (Ostwald 1923). Intensive researches in this area, which began then, resulted in formation of a new science - colloid chemistry. Its creators R. Zigmondi ("he opened access to the world of inaccessible sizes"), T. Svedberg and J. Perrin (for "a breakthrough into the world of discrete particles") were awarded the Nobel Prizes in 1925 and 1926. What is happening in our time is a secondary discovery of the "world of neglected dimensions", now not only at micro, but also at nanoscale.

Mineralogical intervention into the micro and nanoworld began long ago. There is generally nothing revolutionary in what is happening nowadays. Nano-mineralogy is a normal and inevitable stage in the development of mineralogical science. Moreover the role of mineralogy (crystallography) in the study of nanoscale objects is quite comparable with the role of physics, chemistry or biology. Suffice it to recall that structural mineralogy always operated on nanoscale elements, and nucleation and growth of crystals is a typical nanoprocess, a crystalline nucleus is a nanocrystal, and opal is a nanostructured natural material. Nanotechnology is often a repetition of natural processes or nature-like technologies.

The very first results of the mineralogical intervention in the nanoworld were very impressive (Nanomineralogy… 2005). Among the most important achievements of the past years:

– - discovery of a new type of structurally and morphologically ordered objects – nanoindividuals. The likely morphological diversity of nanoindividuals is enormous and not limited by laws of classical crystallography;
– - significant expansion of boundaries of mineral world due to solid amorphous substances previously attributed to mineraloids. A discovery of a new class of structurally ordered mineral structures;
– - finding of the lower limit of mineral objects, beyond which matter is in a different, non-mineral (protomineral, quatarone, cluster) state;
– - identification of common laws of self-organization at nanoscale in the mineral and living worlds. Mineral and living matter, as is known, are formed at nanoscale. Both do not exist outside the lower limit of the nanolevel;
– - substantiation of fundamental role of forms of existence and properties of nanoscale particles in minerals and ores for the development of new technologies for deep and complex processing of mineral raw.

The research of the mineral nanoworld is just beginning. The study of mineral nanostructures, natural clusters, nanostructured natural materials, organo-mineral nanoobjects, nanodispersed phases in ores, development of new technologies for their extraction, modifying properties of minerals and mineral nanoparticles is an incomplete list of the nearest tasks of general and applied nanomineralogy. At that we will have to reconsider a number of fundamental concepts, significantly expand them or introduce new ones. Another argument for the expansion of mineralogical objects to nanoscale is that on this basis a great challenge for mineralogical science can be formed, which is now missing. It is not reasonable to reject a breakthrough promising unprecedented discoveries and a deeper understanding of nature of the mineral matter and its sources. It is hard to deny the intellectual appeal and charm of the nanomineralogical project. The opening door to the nanoworld should not be closed, even if there is a threat of erosion of foundations of classical mineralogy.

Significant progress in understanding properties of the nanoworld was made by the quatarone concept of cluster self-organization of matter developed by us (Askhabov 2011). Within this concept new models of crystal nucleation (Askhabov 2016), formation of various types of nanoparticles (including fullerenes (Askhabov 2005a)) and solid amorphous materials (Askhabov 2005b) are proposed. At that quatarons themselves are new nanoobjects without analogues in the macroworld. They are not small pieces, cut from a large piece of matter or obtained by successive division. They cannot be identified with ordinary clusters—equilibrium structures optimized by geometry or energy.

The quataron concept solved the fundamental problem of deciphering the mechanism of crystal growth in an amazingly simple way (Askhabov 2016). Quatarons proved to be ideal building units for crystal growth. Due to the dynamism of structure, their inclusion into the crystal lattice occurs with virtually no kinetic resistance and deformation of the crystal lattice. The quataron model of crystal growth acts as an alternative to the known models of crystal growth by attaching individual atoms or ready crystalline blocks.

The development of ideas of the quataron concept opened a window to the protomineral world. And this world is today regarded as a new "world of neglected dimensions" (Askhabov 2017). As a result, the program of W. Ostwald receives a new impetus and focuses on the study of objects of the protomineral world. We must answer not only the questions of how minerals are formed, but also why they are formed, why minerals are exactly as they are. The ontogenesis of minerals should begin not with the origin of the mineral, but with the protomineral state of mineral matter.

Thus, the world of minerals is preceded by the world of specific pre-mineral objects —the protomineral world (Fig. 1). This world requires interdisciplinary approaches and new instrumental methods with spatial angstrom-nanometer and temporal femto-nanosecond resolution. In connection with the emergence of the protomineral world in the current agenda of mineralogical science, it is highly desirable to draw up a program of top-priority experiments for implementation with the European free electron laser - a device that is potentially able to satisfy requests of mineralogical science and study in detail the process at the source of mineral matter.

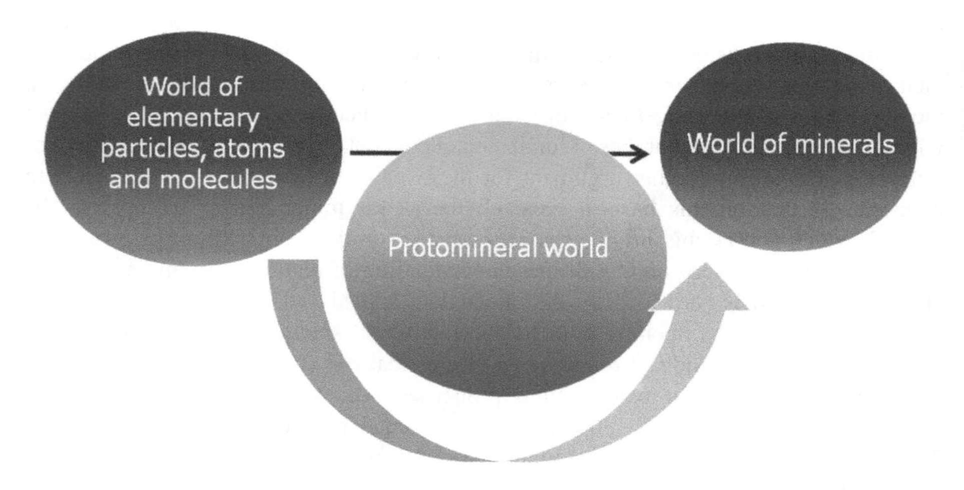

Fig. 1. Way from the world of separate atoms and molecules toward the world of mineral individuals comes through the protomineral world.

Acknowledgements. The work was accomplished with partial financial support by UB RAS program No. 18-5-5-45, and RFBR No. 19-05-00460a.

References

Ostwald W (1923) World of neglected dimensions. Introduction to modern colloid chemistry with an overview of its applications. Mir, Moscow, 228 p (in Russian)

Yushkin NP (2005) Ultra- and microdispersed state of the mineral matter. In: Askhabov AM, Rakin VI (eds) Nanomineralogy. Nauka, St. Petersburg, 581 p (in Russian)

Askhabov AM (2011) Quataron concept: main ideas, some applications. In: Proceedings of komi science center UB RAS.- №3, no 7, pp 70–77 (in Russian)

Askhabov AM (2016) Quataron crystal nucleation and growth models. In: Proceedings of RMS.- Ch. CXLV, no 5, pp 17–24 (in Russian)

Askhabov AM (2005a) The kvataron model of fullerene formation. In: Physics of solid state, vol 47, no 6, pp 1186–1190

Askhabov AM (2005b) Aggregation of quatarons as a formation mechanism of amorphous spherical particles. In: Doklady earth sciences, vol 400, no 1, pp 937–940

Askhabov AM (2017) New stage of mineralogical invasion into the "world of neglected dimensions": discovery of the protomineral world. In: Proceedings of anniversary congress of the Russian mineralogical society "200th anniversary of the Russian mineralogical society", St. Petersburg, vol 2, pp 3–5 (in Russian)

38

Correlation Value of the Mineralogical Composition of Tills

L. Andreicheva(✉)

Institute of Geology KomiSC UB RAS, Syktyvkar, Russia
andreicheva@geo.komisc.ru

Abstract. The material composition of the tills, including its mineral component, is formed during of the exaration-accumulative activity of the cover glacier and depends on the composition of the rocks of the glaciation centers, transit areas, local underlying rocks and on the relief of the preglacial bed. Thus the composition of the till is conditioned by the total influence of source glacier provinces and represents an average sample of the rocks on the way of the glacier.

Keywords: Mineral composition · Correlation · Till · Petrographic composition

1 Introduction

The studies were conducted on a vast and geologically heterogeneous territory of the Timan-Pechora-Vychegda region, which determined the variability of lithological and mineralogical parameters of the tills in term of the regional plan. This restricts their use for correlations. But taking into account the factors of the glacial sedimentation genesis and the complex study of the tills, the stratigraphic confinement of moraine horizons is established very confidently. The features of the mineral-petrographic composition of the tills and the rocks of the source provinces (remote, transit and local) are discussed in detail in the papers by Andreicheva (1994, 2017).

2 Results and Discussion

Pomusovski (Okskii) till in the lower Pechora and Laya is characterized by a low content of a heavy fraction (0.26–0.4%) and amphibole (10–12%) - garnet (12–19%) - epidote (35–41%) mineral association. The concentration of titanium minerals (rutile, titanite, leucoxene) is 7–9%, the total amount of pyrite and siderite does not exceed 10–11%. Eastward (in the valleys of the Kolvaand Pechora Rvers - in borehole 301-Kushshor) the yield of the heavy fraction is significantly higher - 0.8–1.16%, but the epidote content decreases (to 26–30%), amphiboles (to 5–7%) and the total concentration of pyrite and siderite increases sharply to 37%. In borehole 301-Kushshor the number of garnets (up to 23%) and the content of titanium minerals (up to 11%) increase.

The decrease in the content of amphiboles in the Pomusovskitill eastward with the underlying Aptian and Albianrocks. They are characterized by the epidote-amphibole mineral association with a distinct W-E decreasing content of amphiboles. The enrichment of the heavy fraction with pyrite and siderite in the of Kolvavalley is associated with assimilation of these minerals from the Mesozoic sediments. The petrographic composition of the fragments and their south-eastward orientation testifies to the formation of the pomusovskytill in the region due to source from Fennoscandinavia and the Northern Timan.

The composition of the heavy fraction of Pechora(Dneprovian) till is also characterized by areal variability, reflecting the specifics of the mineral composition of the rocks of remote, transit and local source provinces. The minimum yield of the heavy fraction is 0.33%, noted in Pechoratill in the middle Pechora River, the maximum - 0.86%, in the SeydaRver. The mineral association is represented by epidote, garnets, amphiboles, pyrite, siderite and ilmenite. In some areas, the role of titanium minerals increases, most often due to an increase in the amount of leucoxene, sometimes the content of metamorphogenic minerals is increased: kyanite, staurolite, sillimanite.

Epidote prevails in the Pechoratill, its quantity varies from 14 to 37%. Maximum concentrations are recorded in sections of the middle Pechora, the lowest - in the ChernayaRver. The number of garnets is increased to 21% in the ShapkinaRver, reduced to 5% - on the valley of the SeydaRver. The content of amphiboles is low: its maximum concentration (16%) is marked in the VychegdaRver, and the lowest - on the right bank of the UsaRver, where is only 0.6%, which is associated with the complete absence of amphiboles in the underlying Mesozoic rocks. The amount of pyrite also varies significantly: its minimum content (3%) is in the basin of the ShapkinaRver, the maximum in the SeydaRver- 27%. Siderite in the LayaRver is 8%, in the sections in the ChernayaRver reaches of 27%, pyrite and siderite are constantly present. Their total concentrations are different, but the ratio is stable: almost everywhere siderite dominates pyrite. In the petrographic thin sections of the pechorskytill, glauconite is contained in significant quantities (up to 60 grains in a standard petrographic section), which, like pyrite, siderite, is characteristic of local Mesozoic rocks. The composition of the minerals of the heavy fraction testifies to the participation of underlying Triassic, Jurassic and Cretaceous rocks in the formation of the mineral spectrum of the till (Andreicheva and Nikitenko 1989). Single reference boulders are constantly present - pink crinoid-bryozoan Novaya Zemlya limestones. Another characteristic feature of the Pechora glaciation is a steady south-western trend of ice movement, which is consistent with the petrographic composition of rock fragments in the till. The obtained data testify to the Paykhoy-Ural-Novaya Zemlya center of glaciation in the Pechora time throughout the region.

The peculiarity of the paleogeography of Vychegda (Moskovian) glaciation is the existence in the west and in the east of the region of the ice covers of various centers. Accordingly, the material composition of the till was formed by the material of different source provinces. In the western and central parts of the region, the heavy fraction of till, which ranges from 0.42% in the lower Pechora River to 0.95% in the Vychegda valley, is represented by an amphibole-garnet-epidote mineral association with increased total pyrite and siderite contents (up to 32% in the ChernayaRver valley in the north of the Bolshezemelskaya tundra). Sometimes the amounts of ilmenite and titanium minerals are increased. In the south of the region, the association of heavy minerals contains less epidote than in the Pechoratill, and there are more amphiboles and garnets, sometimes amphiboles reach 57% of the heavy fraction. Pyrite and siderite make up the first percentages of the weight of the heavy fraction, and ilmenite concentrations are increased (9-12%). The amount of glauconite in Vychegdatill is 3–4 times lower than in Pechoratill: up to 15–20 grains per standard petrographic section. Heteroagedtills particularly sharply differ from one another by the composition of heavy minerals in the southern regions of the region. The orientation of the fragments in sector 270–3600 and the presence of indicator rocks of the North-West Terrigenous Mineralogical Province also indicate the migration of material from the west and from the north-west. Another feature of this till is the predominance of light colored limestones in the group of carbonate rocks.

In the northeast of the region, the mineral composition of the heavy fraction of the Vychegdatill is variable and consists of ilmenite (12%), garnet (15%), pyrite (15%), siderite (25%), epidote (32%). The content of titanium minerals is slightly increased – 10%, and amphiboles – is decreased (4–12%). The petrographic spectrum of fragments represents Uralian rocks. The fragments of rocks are oriented in sub latitudinal and SW directions, i.e. material for the formation of Vychegdatill came from the east-northeast.

The main mineral association of the Polar (Ostashkovian) till contains epidote (19–27%), garnets (14–20%) and amphibole (11–16%). The total content of siderite and pyrite is 13–35% with dominating of siderite. The crinoid-bryozoanNovaya Zemlya limestones are present. The orientation of detrital material from NNE to SSW confirms the onset of the cover glacier in polar time from Pay-Khoy-Novaya Zemlya.

3 Conclusions

Despite the variability of lithologic and mineralogical parameters of tills in terms of the regional plan, the data of the mineral composition of tills can be used with confidence for the dismemberment of Quaternary sections and wide spatial correlations of glacial deposits, taking into account the factors of glacial sedimentation genesis, as well as the petrographic composition of rock fragments and their orientation.

Acknowledgements. This research was supported by UB RAS project № 18-5-5-50 and project AAAA-A17-117121140081-7.

References

Andreicheva LN (2017) Correlation of neopleistocene tills of the north of the Russian plain in petrographic composition. Lithol Miner (1):82–94. (in Russian)

Andreicheva LN (1994) The source provinces and their influence on the formation of the composition of the moraines of the Timan-Pechora-Vychegda region. Lithol Miner (1):127–131. (in Russian)
Andreicheva LN, Nikitenko IP (1989) The mineral composition of the fine earths of the main moraines of the Timan-Pechora-Vychegda region. In: Mineralogy of the Timan-Northern Ural region, Syktyvkar, proceedings of the institute of geology, Komi SC UB RAS, no 72, pp 52–62. (in Russian)

39

High-Tech Elements in Minerals of Massive Sulfide Deposits: LA-ICP-MS Data

V. Maslennikov[1(✉)], S. Maslennikova[1], N. Aupova[1], A. Tseluyko[1], R. Large[2], L. Danyushevsky[2], and U. Yatimov[1]

[1] Institute of Mineralogy, Ural Branch of RAS, Miass, Russia
mas@mineralogy.ru
[2] CODES ARC Centre of Excellence in Ore Deposits, University of Tasmania, Hobart, Australia

Abstract. LA-ICP-MS data on trace element zonation reflecting a local variation of physicochemical conditions of mineralization in black, grey and clear smokers from the Pacific and Atlantic oceans are used for comparison with the ancient chimneys of the Urals, Rudniy Altai, Pontides and Hokuroko massive sulfide deposits. Host rocks also influence on high-tech trace element assemblages in chalcopyrite: ultramafic (high Se, Sn, Co, Ni, Ag and Au) → mafic (high Co, Se, Mo and low Bi, Au and Pb) → bimodal mafic (high Te, Au, Ag, Bi, Pb, Co, moderate Se, and variable As and Sb) → bimodal felsic (high As, Sb, Mo, Pb, moderate Bi, and low Co, Te and Se). In sphalerite of the same range, the contents of Bi, Ga, Pb, In, Ag, Au and Sb increase versus Fe, Se, Sn and Co. The variations in pyrite coincide with these changes. Diagenetic evolution of high-tech elements is recognized in sulfide nodules.

Keywords: LAICP-MS · Trace elements · Chimneys · Massive sulfide deposits

1 Introduction

The bulk minor and trace element analyses indicate significant influence of host rocks on the metal inventory of the modern and ancient massive sulfide deposits. However, these general data can't provide dramatic improvement of the extraction of high-tech elements from massive sulphide ores. In contrast, the results of the quantitative laser ablation inductively coupled plasma mass spectrometry (LA-ICP-MS) show concentration of most trace metals in pyrite, chalcopyrite and sphalerite varieties as a function of physicochemical parameters of the fluid as well as host rock composition. Local LA-ICP-MS analyses of the sulphides provides a new insight into the processes of ore treatment and recovery of high-tech elements. The chimney fragments, sulfide turbidites, and their ore diagenites are suitable subject of this style of the research. The study is a key basement for fither improvement of mining and metallurgical processes.

2 Methods and Approaches

In this paper, we use new LA-ICP-MS data on minor and trace elements in sulfides from Rudniy Altai, Pontides and Hokuroko massive sulfide deposits. Original LA-ICP-MS data on smokers from the Pacific and Atlantic oceans are used for comparison with the ancient chimney material. The sulfide nodules are the other important subject of the research reflecting chemical evolution of diagenetic and metamorphic processes.

Quantitative LA-ICP-MS analysis of sulfides from chimneys for a wide range of major and trace elements (Fe, Cu, Zn, Co, Ni, Au, Ag, Bi, Pb, Tl, Cd, As, Te, Se, Mo, Sn, V, Ti and Mn) was carried out using a New Wave 213 nm solid-state laser microprobes coupled to an Agilent 4500 quadrupole ICP-MS housed at the CODES LA-ICP-MS analytical facility, University of Tasmania and Institute of Mineralogy UB RAS. For this study, quantitative analyses were performed by ablating spots ranging in diameter from 40 to 60 microns. Data reduction was undertaken according to standard methods (Danyushevsky et al. 2011). The LA-ICP-MS was also used for detection of rare mineral microinclusions in the sulfides studied and for analysis of trace elements in telluride phases. Each zone of the chimney and nodules are characterized by LA-ICP-MS-images (Large et al. 2009).

3 Results and Discussion

A variety of very well preserved modern black, grey, white and clear smokers (Pacific and Atlantic oceans) and fossil vent chimneys from the Urals, Rudniy Altai, Pontides and Hokuroko massive sulfide deposit show systematic trace element distribution patterns across chimneys. Chalcopyrite is enriched in Se, Sn, Bi, Co, Mo, and Te. Sphalerite in the conduits and the outer chimney wall contains elevated Sb, As, Pb, Co, Mn, U, and V. He highest concentrations of most trace elements are found in colloform pyrite within the outer wall of the chimneys, and likely result from rapid precipitation in high temperature-gradient conditions. The trace element concentration in the outer wall colloform pyrite decrease in the following order, from the outer wall inwards; Tl > Ag > Ni > Mn > Co > As > Mo > Pb > Ba > V > Te > Sb > U > Au > Se > Sn > Bi, governed by the strong temperature gradient. In contrast, pyrite in the high- to mid-temperature central conduits exhibit concentration of Se, Sn, Bi, Te, and Au. The zone between the inner conduit and outer wall is characterised by recrystallization of colloform pyrite to euhedral pyrite, which becomes depleted in all trace elements except Co, As and Se. The mineralogical and trace element variations between chimneys are likely due to increasing fO_2 and decreasing temperature caused by mixing of hydrothermal fluids with cold oxygenated seawater (Maslennikov et al. 2009).

Host rocks also influence on high-tech trace element assemblages in chalcopyrite: ultramafic (high Se, Sn, Co, Ni, Ag and Au) → mafic (high Co, Se, Mo and low Bi, Au and Pb) → bimodal mafic (high Te, Au, Ag, Bi, Pb, Co, moderate Se, and variable As and Sb) → bimodal felsic (high As, Sb, Mo, Pb, moderate Bi, and low Co, Te and Se). In sphalerite of the same range, the contents of Bi, Ga, Pb, In, Ag, Au and Sb increase versus Fe, Se, Sn and Co. The variations in pyrite coincide with these changes (Maslennikov et al. 2017).

The next stage of mineral and chemical differentiation is halmyrolysis of clastic sulfide sediments followed by diagenesis, anadiagenesis and metamorphism. Halmyrolysis as a leaching process of clastic sulfide sediments give way to redeposition of high-tech element such as Se, Te, Au, Ag, Sn, and In as own minerals in enrichment submarine supergene zones. The inclusions of authigenic mineral such as Pb, Ag, Au-tellurides, Ag and Pb-selenides, In, Cu, Ag, Sn, Ge-sulfides, and Sn-oxide are increased in sizes in the enrichment submarine supergene zones of massive sulfide deposits.

In many massive sulfide deposits, sulphide nodules have a zonal structure: the nucleus of the poikilite pyrite (zone A) is successively surrounded by the fringes of metacrystalline pyrite (zone B) and marcasite (zone C). Each zone is characterized by own mineralogical features, which are reflected in the results of LA-ICP-MS-images ("microtopochemistry") of the surface of the nodule cut. It is assumed that at the first stage, the source of the substance for nodule growth was the products of the halmyrolysis of hyaloclastites containing an admixture of sulphide material. In the diagenetic nucleus of nodule, the chemical elements typical of illite (Si, Al, K, Mg, V, Cr,), rutile (Ti, W), apatite (Ca, Mn, U), galena (Pb, Bi, Sb, Ag), bismuth sulphosalts (Bi, Cu, Pb), bornite (Cu, Bi, Ag), tetrahedrite-tennantite (Cu, As, Sb), chalcopyrite (Se, Te, Cu), native gold (Au, Ag, Hg) and barite poikilites. In the pyrite, the high contents of Co and Ni are suggested to be substituted for Fe^{2+}. Only signs of tellurides of bismuth are guessed. Zone B, probably formed in the early anadiagenetic stage, is depleted of most trace elements, with the exception of Cu, Pb, and Ag. Formation of latest zone C was accompanied by of Cd, In, Tl, As, Sb, Mo, and Ni saturation.

4 Conclusions

The mineralogical and trace element variations between chimneys are likely due to increasing of O_2 and decreasing temperature caused by mixing of hydrothermal fluids with cold oxygenated seawater. The next stage of trace element differentiation is halmyrolysis of sulfide sediments followed by diagenesis, anadiagenesis and metamorphism. Halmyrolysis as a leaching process of clastic sulfide sediments gives a way to redeposition of high-tech element such as Se, Te, Au, Ag, Sn, and In as own minerals in enrichment submarine supergene zones. The enrichment in the core of sulfide nodules is changed to trace element depletion in later epigenetic stages.

Local LA-ICP-MS analyses of the sulphides provides new insight into the processes of selective ore treatment and recovery of high-tech elements from different genetic ore varieties in implication to diverse ore formational types of massive sulfide deposits.

Aknowledgements. LA-ICP-MS analyses were carried out during visiting programs (2005, 2009, 2012, 2013, 2015) sponsored by the ARC Centre of Excellence grant to CODES. The research of sulphide nodules was supported by the Russian Foundation for Basic Research (project no. 17-05-00854) in the Institute of Mineralogy UB RAS.

References

Danyushevsky LV, Robinson R, Gilbert S, Norman M, Large R, McGoldrick P, Shelley JMG (2011) Routine quantitative multi-element analysis of sulfide minerals by laser ablation ICP-MS: standard development and consideration of matrix effects. Geochim Explor Environ Anal 11:51–60

Large RR, Danyushevsky L, Hillit H, Maslennikov V, Meffere S, Gilbert S, Bull S, Scott R, Emsbo P, Thomas H, Singh B, Foster J (2009) Gold and trace element zonation in pyrite using a laser imaging technique: implications for the timing of gold in orogenic and Carlin-style sediment-hosted deposits. Econ Geol 104:635–668

Maslennikov VV, Maslennikova SP, Large RR, Danyushevsky LV (2009) Study of trace element zonation in vent chimneys from the Silurian Yaman-Kasy VHMS (the Southern Urals, Russia) using laser ablation inductively coupled plasma mass spectrometry (LA-ICP MS). Econ Geol 104:1111–1141

Maslennikov VV, Maslennikova SP, Ayupova NR, Zaykov VV, Tseluyko AS, Melekestseva IY, Large RR, Danyushevsky LV, Herrington RJ, Lein AT, Tessalina SG (2017) Chimneys in Paleozoic massive sulfide mounds of the Urals VMS deposits: mineral and trace element comparison with modern black, grey, white and clear smokers. Ore Geol Rev 85:64–106

40

Crystallomorphology of Cassiterite and its Practical Importance

I. Vdovina(✉)
Nizhny Novgorod Institute of Education Development,
Nizhny Novgorod, Russia
viann@inbox.ru

Abstract. The methods of applied mineralogy are used to assess the scale of the mineralization of tin occurrences and deposits. One of such methods is the crystallomorphological method for estimating the level of tin ore bodies developed by N.Z. Evzikova. According to the crystallomorphological features of cassiterite it allows one to estimate the possible depth of distribution of mineralization and its magnitude. Using the example of three studied tin-ore districts of the Far East the possibilities of using crystallomorphology of cassiterite to predict and evaluate the prospect of mineralization of both the ore region as a whole and individual ore occurrence at the exploration stage are shown.

Keywords: Applied mineralogy · Crystallomorphology · Crystallomorphological method · Cassiterite · Scale of mineralization · An assessment of prospects

1 Introduction

The task of screening ore occurrences has always been and remains one of the important tasks of the geological survey. One of the achievements of applied mineralogy in the history of its development is participation in metallogenic researches and the development of the theory of mineral deposits prediction. Among the tasks solved by the methods of applied mineralogy (in particular the search for crystallomorphology) there are tasks aimed at determining the scale of mineralization (its vertical extent and the magnitude of the erosion-denudation slice). The methods of the search crystal morphology are based on two important postulates: the ontogenic development of the mineral form and the doctrine of typomorphism. One of such methods is crystallomorphological method for estimating of tin ore occurrences developed by N. Z. Evzikova on the example of tin deposits.

2 Methods and Approaches

Crystallomorphological method for evaluating of tin ore occurrences (Evzikova 1984) is based on the difference in the natural appearance of cassiterite crystals in different parts of the ore-bearing system as they move away from the source. The deeper parts of the ore bodies correspond to short-prismatic, isometric forms of cassiterite (I and II types), the higher—elongated forms of crystals (IV and V). The scale of the mineralization and the degree of its preservation reflect the crystallomorphological criteria: scoring represented by the statistical predominance of types IV and V and values from 50 to 200; degree of elongation of crystals (object's class) reflecting the completeness of the crystallomorphological evolution; the homogeneity of the morphology of cassiterite in different height sections of the ore object. The scoring factor is calculated by the formula $B = [2\ (V) + (IV)] - [2\ (I) + (II)]$, where I, II, IV, V is the percentage of crystal morphological types in the sample.

The testing of this method at various times took place at tin ore deposits of the Far East where for the first time this method was used to evaluate the prospects for tin ore mineralization in exploration. The author has researched the crystallomorphological features of cassiterite from the ore manifestations of three Far Eastern tin ore regions (the Komsomolsk, Badzhal, Yam-Alin regions) in order to obtain a crystallomorphological estimate of the prospect of their mineralization.

3 Results and Discussion

Ore districts are located within the Khingan-Okhotsk volcano-plutonic belt and the same-name tin-bearing region. The Khingan-Okhotsk volcano-plutonic belt represents a vast magmatic area of the northeast strike superimposed on a heterogeneous foundation formed in the atmosphere of the transform continental margin. The Badzhal region is located in the junction zone of ancient Archean-Proterozoic cratons and Mesozoic accretionary folded belts and is characterized by the presence of a Cretaceous volcano-tectonic formation in the central part of the granitoid batholith.

Komsomolsk and Yam-Alin regions are located in accretionary folded systems spatially coincides with the eponymous volcanic zone (Rodionov 2005).

The Komsomolsk Ore Region. Several hundreds of ore mineralization zones, mainly cassiterite-silicate formation, have been established within the Komsomolsk ore region. By the beginning of 90s of last century crystallomorphological researches covered about 40 manifestations of tin mineralization in which a fairly representative number of samples were analyzed. 25 ore manifestations of this number were evaluated at deep horizons in the process of prospecting and evaluation (14) or exploration which made it possible to determine the reliability of forecasts obtained earlier by the crystallomorphological method.

When sorting out 25 ore occurrences by crystallomorphological method 11 mineralization zones received a positive forecast. Subsequent testing at the exploration stage confirmed this forecast for 6 ore occurrences. In five positively evaluated by the subsequent preliminary exploration the forecast was not confirmed.

Among the ore occurrences that received a negative assessment on the depth 10 objects were studied. The convergence of the results in these cases turned out to be higher. A negative forecast was confirmed by 8 occurrences. And only in two zones contrary to expectations industrial ores have been established.

The Badzhal Ore Region. The crystallomorphological researching of cassiterite from the ore occurrences of the Badzhal ore district began with the first explorations in the middle of 70s of the last century (Vdovina 2005).

In the process of research at the exploratory stage the cassiterite ore occurrences of the "Badzhal axis" structure were studied. Crystallomorphological criteria indicated a high level of slice and insignificant prospects for ore occurrences. No detailed work has been done.

The Pravourmiyskoe deposit was the most studied in details. The morphology of cassiterite crystals of this deposit was researched at all stages of prospecting and exploration (Vdovina 1987, 2008).

At the first stage of work the crystallomorphological mapping was carried out on surface mine workings. Evaluation of the mineralization's prospect was as follows:

- the level of the denudation slice is high enough;
- the gradient of the vertical crystallomorphological variability is large; the deposit is characterized by the proximity of the upper and lower boundaries of mineralization;
- the estimated length of the mineralization to the depth is 200–220 m;
- the morphology of the crystals suggests a large length of mineralization to the depth and the occurence of industrial ores in the middle part of the deposit.

Material was obtained from the deep horizons of the deposit in the detailed exploration. The results of the crystallomorphological analysis of cassiterite of the Pravourmiyskoe deposit were confirmed by the results of exploration. The length of the mineralization to a depth was no more than 250 m. The level of the denudation slice is really high. In addition the results of cassiterite of the Pravourmiyskoye deposit's crystallomorphological features researching confirmed the regularity of the spatial and temporal distribution of the morphological types of cassiterite. The same sequence of the cassiterite crystals changing from type I to type V was established at the stockwork type of the cassiterite-quartz formation. The deposit is currently being exploited.

About a dozen have been studied *within the Verkhnebadzhalsky ore cluster*. Most detailed and from the surface and from the depth Blizhnee deposit has been explored (Vdovina 2011).

The Blizhnee Deposit. Cassiterite is of a short-prismatic almost isometric habit, IV and less than V type, the score is 92–130. According to the crystallomorphological features the prospect of mineralization to a depth is small. Exploration data did not confirm the depth of the deposit.

The Talidzhak mine is the most perspective one. Cassiterite is predominantly IV and V types, of medium degree of elongation, the scaling varies within 100–200. The gradient of crystal morphological variability is weak. Ore occurrences are characterized as poorly eroded. The vertical extent of mineralization is assumed to be 500 m. The detailed exploration has not been carried out.

The Yam-Alin Ore Region. Ore occurrences were researched only at the exploration stage. There is the Sorukan deposit. Cassiterite is situated in the zones of quartz-chlorite composition in the form of scattered and nested impregnation and quartz-cassiterite lenses and veinlets. Pyramid-prismatic crystals of type IV of crystallomorphological type prevail in their composition, crystals of type V are found in a small amount. There are a lot of prismatic fragments of crystals of very different lengths including conical. All type IV crystals are characterized by a moderate and strong degree of elongation, a zonal structure; growth zones fix the development of cassiterite in types II–IV. The deposit's class development is 4. The deposit is poorly eroded and perspective on depth.

The ore occurrences of the Shiroky, Exan and Bastion sections from the different ones are of interest. Cassiterite is very small, almost colorless, slightly yellowish, often observed in star splices and aggregates. It is presented by IV and V crystallomorphological types with a predominance of type V. The crystals are characterized by short heads (dull pyramid), moderate and strong degree of elongation. The crystals are transparent, zonal; the observed growth phantoms show their development in type IV and III. Often there are simple articulated twins. The presence of cassiterite of different physical and crystallomorphological properties gives the right to assume the development of tin mineralization in several stages. In general the district is prospective, poorly eroded.

According to the results of the crystallomorphological method the most promising sections include the majority of the studied sections.

4 Conclusions

The author believes that the use of crystallomorphological analysis to determine the denudation slice, the extent of mineralization to the depth and generally for the industrial evaluation of tin mineralization is appropriate at the exploration stage. But the credibility and reliability of the assessment will increase if it is used in combination with others, in particular, at the first stage with structural and morphostructural (Vdovina 2008), and at later ones with more "direct" mineralogical, geochemical and other researches.

References

Evzikova NZ (1984) Prospecting cristal morphology, Moscow

Rodionov SM (2005) Metallogeny of tin of East of Russia, Moscow

Vdovina IA (1987) Crystallomorphology of cassiterite as one of the criteria for industrial evaluation of the tin ore deposit in notes all-union. Mineral Soc Part XVI, I:60–65

Vdovina IA (2005) Crystallomorphological features of cassiterite occurrences in the Badzhal ore region. Nat Geogr Res, Komsomolsk-na-Amure, pp 4–13

Vdovina IA (2008) Morphostructural-crystallomorphological evaluation of the prospect of mineralization (on the example of the Badzhal tin ore region) in Fedorov session 2008. Thesises of international scientific conferences, Saint-Petersburg, pp 246–249

Vdovina IA (2011) About the crystallomorphology of cassiterite. In: X international conference «New ideas in the Earth sciences». Reports in 3 volume. Russian State Geological Exploration University, Moscow, vol 1, pp 107–108

Industrial Minerals, Precious Stones, Ores and Mining

41

Shungites and their Industrial Potential

V. Kovalevski and V. Shchiptsov(✉)

Institute of Geology, Karelian Research Centre, Petrozavodsk, Russia
shchipts@krc.karelia.ru

Abstract. Shungite rocks are widespread in Zaonezhye, Republic of Karelia, where they constitute dozens of carbonaceous rock deposits of the Paleoproterozoic Onega structure with predicted carbon resources of more 4 billion tons. The lower age boundary is of 2.1 Ga. Shungite rocks belong to carbonaceous rock class. These rocks metamorphosed in greenshcist facies of muscovite-chlorite-biotite subfacies are unique natural, noncrystalline, non-graphitized, fullerene-like carbon. They have various structural-mineralogical levels: (a) supramolecular, (b) molecular, (c) electron-energetic, (d) structural-physical and (e) geologic-genetic (parametric). Shungite rocks contain shungite carbon (shungite matter) and a variety minerals, microminerals and nanominerals. The applications of shungite rocks are determined with regard for their natural types. Authors had shown their intergrated application in ore-thermal processes.

Keywords: Shungite rocks · Carbon · Metamorphism · Paleoproterozoic · Application

Shungite rocks, named so after a Karelian town, Shunga, have attracted attention for decades and have no counterparts in the Earth's geological evolution with respect to the mode of occurrence and tremendous reserves. They are unique from both scientific and practical points of view. Shungite rocks are part of Paleoproterozoic carbonaceous formation in Karelia. They are common in the Trans-Onega area, Russian Karelia, where 25×10^{10} tonnes of autochthonous organic matter have been formed over an area of about 9000 km^2. These complexes constitute dozens of carbonaceous rock deposits in the Paleoproterozoic Onega Structure with forecast carbon reserves of over 4 billion tones and are mainly confined to the rocks of the Ludicovian system with the lower age boundary of 2.1 Ga. The bulk of free carbon (Corg.) is in the Trans-Onega suite of the Ludicovian superhorizon. Phenomenal carbon accumulation in this superhorizon is responsible for the mineralogenic specialization of the rocks of the Trans-Onega suite. Their Corg. concentration varies from less than 1% to 70 wt% in rocks and up to 98 wt% in anthraxolite aggregates. Carbon concentration in the rocks of the Kondopoga suite of the Kalevian superhorizon is not more than a few percent. Thus, the black shale formation of the Onega Structure is formed of the rocks of the Trans-Onega, Suisari and Kondopoga suites. The world's largest Zazhogino Ore Field, covering an area of 3240 km^2, with two active quarries, Zazhogino and Maksovo, has been identified and relevant evidence was presented (Table 1).

Metamorphism and metasomatism have contributed markedly to the genetically distinctive shungite rocks. They belong to a metamorphogenetic class of shungite formation. Their industrial properties are controlled by metamorphism and a special

Table 1. Comparative description of the parameters of major deposits in Zazhogino Ore Field

Ore body shape	Size		Maximum thickness	Average free carbon content, %
	Length, m	Width, m		
Shunga deposit				
Sheet-like	1400	300	5.2	41
Maksovo deposit				
Sheet-like-cone shaped	700	500	120	40
Zazhogino deposit				
Sheet-like-cone shaped	400	300	60	27
Zalebyazhskoe deposit				
Sheet-like	2000	700	38	35

contribution of K-Na alkaline metasomatism which has affected the formation of the natural types and varieties of shungite and, correspondingly, industrial types and varieties and applications of this unique raw material. On the P-T scheme, showing principal correlations between metamorphic facies and subfacies after S.A. Bushmin and V.A. Glebovitsky, shungite rocks were derived under greenschist-facies, muscovite-chlorite-biotite-subfacies conditions of metamorphism at a temperature of 325–450 °C and a pressure of 2–5 kbar (Bushmin and Glebovitsky 2016).

The mineral form of shungite (shungite matter) is non-graphitizable fullerene-like carbon, which differs from graphite at a supramolecular, atomic and zonal (electron) structure level. The main supramolar character of shungite is an ability to form spherical structures (empty globules). At an atomic level, in addition to hexagonal rings alone typical of graphite, pentagonal and heptagonal rings, characteristic of fullerene-like structures, are also observed. At a zonal structure level, the energy of collective excitations of valent (external) and framework (internal) π- and σ-electrons decreases relative to graphite, which is typical of fullerenes as well (Kovalevsky et al. 2016). Shungite from some deposits displays diamagnetic properties characteristic of fullerenes. The structure of shungite rocks is similar to that of vitreous-crystalline materials, where highly dispersed crystals are distributed in non-crystalline matrix.

Shungite rocks belong to a class of carbonaceous rocks varying in carbon content and mineral diversity and those are natural carbon-mineral composite materials. They contain shungite (shungite matter) and a variety of micro- and nanominerals. Silicate minerals are highly disperse and are evenly distributed in carbon matrix. Major rock-forming minerals are quartz, mica, albite and pyrite. High secondary and accessory mineral concentrations and a certain spectrum of layered and cluster impurities are observed. Natural types of shungite rocks display several textural and structural varieties and the non-uniform phase composition of carbon and geochemical characteristics.

Shungite applications are determined by natural types and varieties of shungite rocks (Kovalevsky et al. 2016). The structure and properties of shungite rocks are responsible for their application in oxidation-reduction processes: in blast furnace production of foundry (high-silicon) cast iron: in ferroalloy production; in yellow phosphorus production; in carbide and silicon nitride production; as a reinforcing component of groove masses; as a filler of non-stick paints.

The sorptive catalytical and reduction properties of shungite rocks are used: for treatment of high-quality drinking water in flow systems and wells; for removal of many contaminants from urban domestic and industrial sewage; for swimming-pool water treatment; for water treatment at heat power plants; for electrically conductive paint production; for electrically conductive concrete and brick production; for electrically conductive and plastering and masonry solutions; for electrically conductive asphalt production. These materials were used for developing and designing heaters: rooms screening electromagnetic radiation; a method for removal of ice from roads (warm pavements and roads).

Finely ground shungite can be mixed with any binders of organic and inorganic origin and can thus be used as: black pigment for oil and water paints; a filler for polymers materials (polyethylene, polypropylene, fluoroplastic, etc.); substitute for white soot and technical carbon in rubber production.

One essential feature of shungite rocks is that they can be modified with regards for their desirable application. Enrichment of shungite rocks and their division into micro- and nanosized components make a possible to activate shungite carbon and expand the potential applications of shungite rocks in science-intensive technologies, e.g. nanotechnologies.

There is only one geologo-industrial classification of shungite rocks (after Yu. Kalinin) based on industrial types distinguished with respect to their mineralogical composition and corresponding applications. However, the latest results of the large-scale practical application of shungite rocks shows that such a division into industrial types is clearly insufficient. Therefore, division into subtypes or varieties is required.

At the present stage of the study of Karelia's shungite rocks the goal of research and appraisal is not only to estimate the reserves of potential deposits and to assess the petrographic and structural-chemical characteristics of shungite rocks but to develop criteria and recommendations for the industrial application of rocks from potential deposits in innovative and science-intensive fields (Kalinin and Kovalevski 2013).

For this purpose it is proposed:

1. To re-assess earlier geological, physico-chemical and technological data on known deposits to make up a list of known shungite rocks suitable for particular applications.
2. To reveal promising shungite rock applications (e.g. metallurgy, tyre production, water disposal, the production of composite materials, etc.) and certification requirements to be met by raw material for each application.
3. To specify the critical properties of shungite rocks for each application (e.g. chemical composition, mineralogical composition, the structural parameters of carbon and rocks) and express methods for their analysis.
4. To develop certification requirements for shungite rocks with regard for their applications.

5. To analyze shungite rock samples and cores from each of the bodies and rock exposures and classify all shungite rocks into types and subtypes referenced to a major application, based on the criteria developed.
6. To correlate the geological parameters of the deposits and outcrops with the geochemical and structural-petrological characteristics of shungite rocks to forecast the prospecting of shungite rocks with preset properties for required applications.

A classification of the geologo-industrial types of shungite rocks will be worked out and the most promising shungite rock prospects for each application will be specified, based on the results of prospecting and prospecting-and-appraisal for shungite rocks. The accomplishment of the work planned will result in the efficient investment of the money spent in the cost of future deposits and the updating of innovative approaches to the use of Russia's unique carbonaceous raw material.

Shungite has unveiled many of his mysteries, has become known all over the world, has attracted the interest of experts by its great potential and yet has remained largely unknown and open to new discoveries.

Acknowledgements. The work is performed in the framework of the PFNI GAN research of IG KarRC RAS.

References

Bushmin SA, Glebovitsky VA (2016) Scheme of mineral facies of metamorphic rocks and its application to the Fennoscandian Shield with representative sites of orogenic gold mineralizaion. Trans KRC RAS 3–27

Kalinin Yu, Kovalevski V (2013) Shungite rocks: scientific search horizons. Nauka v Rossii 6:66–72

Kovalevsky V, Shchiptsov V, Sadovnichy R (2016) Unique natural carbon deposits of shungite rocks of Zazhogino ore field, Republic of Karelia, Russia. In: SGEM Conference Proceedings of International Multidisciplinary Scientific GeoConference Surveying Geology and Mining Ecology Management, pp 673–680

42

Peridot: Types of Deposits and Formation Conditions

S. Sokolov(✉)

FSBE "All-Russian Institute of Mineral Raw Materials" (VIMS), Moscow, Russia
vims-sokol@mail.ru

Abstract. The generalized information about the peridot from deposits and ore occurrence of well-known formation type is presented in this paper. Peridots from specific deposits vary in the accompanying mineral associations. These gems, depending on the genesis, contain differing in phase composition inclusions of mineral-forming medium and have different crystallization temperatures.

Keywords: Peridot · Formations · Deposits · Origin · Temperature of formation

1 Introduction

The increased interest to peridot – jewelry varieties of olivine, was observed over the last years. This has resulted in numerous publications on the geology, mineralogy, and genesis of this gem. Peridot has been found in almost 30 countries and on all continents, including Antarctica (gem-quality mineral is present in the basalts from the Ross Island); furthermore it was diagnosed in several stony-iron meteorites. Peridot is present in different geological formations, which almost all (with the exception of exogenous placers) belong to derivatives of endogenous processes and often genetically related to ultrabasic and basic rocks.

2 Formation Types of the Deposits of Peridot

Diamond-bearing kimberlites, concentrating peridot in different quantities, are known in Yakutia (Udachnaya-Vostochnaya and Mir pipes), the Republic of South Africa (Kimberly and De Beers pipes), and in Tanzania (Mvadun pipe).

Ultrabasic-alkaline rocks and carbonatites – UAC complexes. Peridot installed at complexes of the Kola peninsula (Kovdor) and Polar Siberia (Kugda).

Basalts and basaltoids (normal and (sub) alkaline). Peridot phenocrysts in the nodules of peridotites and olivenites are situated in lava flows and volcanic craters at deposits and ore occurrence in Russia, USA, Hawaiian Island, the Czech Republic, Madagascar Island, Ethiopia, China, Pakistan, Australia, Antarctica.

Alpinotype hyperbasites. The most famous deposit of peridot – Zabargad Island is classed to this formation type. They are known also in Russia at the East Sayan and at the South Urals.

Placer deposits. Eluvial-deluvial plasers are the most productive among the exogenous deposits of peridot. They are connected with different genetic types of primary deposits of this gem (Zabargad, San-Carlos, Kugda).

3 Origin and Formation Conditions of Peridot

Peridots from the deposits of different formation clearly differ in the associations of accompanying minerals.

A detailed study of the jewelry peridots from the Kovdor and Kugda deposits allowed us to determine the specific features of their micromineralogy, (Yarmishko and Sokolov 2005): crystalline inclusions, represented mainly by minerals accompanying to peridot; decomposition products (magnetite dendrites and magnetite-diopside plates; crystallized melt inclusions.

Fixation multiphase crystallized melt inclusions in peridots confirms the magmatic nature of the deposits some formation types. For example, olivine and calcite from kimberlites of the Udachnaya-Vostochnaya pipe contain the melt inclusions the total temperature range of homogenization of which is 1100–880 °C (Tomilenko et al. 2009). Peridots from basalts of the Hawaiian Islands contain the inclusion of glass (Gubelin and Koivula 1992). The melt inclusions in jewelry peridot from the Kovdor deposits were homogenized at temperatures of 970–930 °C. Crystallization of peridot at the Zabargad deposit occured in temperature range from 900 to 750 °C (Maaskant 1986).

References

Gűbelin EJ, Koivula JI (1992) Photoatlas of inclusions in gemstones. ABC Edition, Zurich

Maaskant P (1986) Electron probe microanalyses of unopened fluid inclusions, semiquantitative approach. Neues Jahrb Miner 7:297–304

Tomilenko AA, Kovyazin SV, Dunlyansky YuV, Pokhilenko LN (2009) Primary melt and fluid inclusions in minerals from kimberlites of the Udachnaya-Vostochnaya pipe, Yakutia. ECROFI-XX. Abstracts, University of Granada, pp 255–256

Yarmishko SA, Sokolov SV (2005) Micromineralogy of peridots from rocks of alkaline-ultrabasic massifs. In: VII International Conference «New Ideas in Earth Sciences». Abstracts, vol 2, Moscow, p 76. (in Russian)

43

Gold and Platinum Group Minerals (PGM) from the Placers of Northwest Kuznetsk Alatau (NWKA)

V. Gusev[1,2(✉)], S. Zhmodik[1,2], G. Nesterenko[2], and D. Belyanin[1,2]

[1] Department of Geology and Geophysics, Novosibirsk State University, Novosibirsk, Russia
vityansky@igm.nsc.ru
[2] Institute of Geology and Mineralogy SB RAS (IGM SB RAS), Novosibirsk, Russia

Abstract. Native gold and PGM from NW Kuznetsk Alatau (NWKA) (South Siberia, Russia) have been investigated. Applying the complex of advanced analytical, geo-chemical, and statistical techniques permits determination of motherlode types of noble mineralization (NM). For gold, it is mineralization of three types: 1 – the gold-sulfide-quartz type associated with dykes of the basic composition and fault-line zones (2), as well as the gold-skarn type (3). For PGM it is mineralization of the Ural-Alaskan (1) and the ophiolitic (2) types, as well as multicomponent alloys associated with layered intrusions (3). The presence of rims and inclusions is indicative of postmagmatic transformations of the minerals. In the meantime, Au and PGM from NW Kuznetsk Alatau placers retain genetic traits of motherlodes.

Keywords: Alluvial placers · Gold · Platinum Group Elements (PGE) · Altai–Sayan folded area · Kuznetsk Alatau

1 Introduction

Au placers with PGM (0.1–0.2% and more by weight of Au) are widespread on the territory of the Kelbes placer region NWKA. The placers were abandoned during last centuries, but motherlode deposits have not been discovered here. Au-quartz, Au-quartz-sulfide types, and Au-magnetite (skarn) have been discovered in NE and the central Kuznetsk Alatau region, as well as in Salair range, and PGE have been revealed in chromites and dykes of the basic composition. The contribution presents new data on the morphology, micro-inclusions, change types, and the composition of native Au and PGM from placers and their comparison with NM from motherlodes of adjacent regions in the South Siberia.

2 Methods and Approaches

Minerals were selected from heavy mineral concentrates under binocular microscope. An Axio Scope A1 (Carl Zeiss) microscope was used for determination of sizes and morphology of NM particles, and then polished preparations were made from these particles for microscopic investigations by methods of ore and electron microscopy. The composition and interrelations between minerals, micro-inclusions, and newly formed phases were studied by SEM (MIRA 3 LMU, Tescan Ltd.), and EMP methods (Camebax Micro) in the share use Center for Multielement Isotope Studies (Novosibirsk).

3 Results and Discussion

The examined gold has different morphologies (sizes, degrees of grain rounding, and deformation), it is characterized by the wide range of variations in the fineness (from 720‰ to 1000‰), by different degrees of chemical change, contains inclusions of quartz, magnetite, and clay minerals, and it is coated with material consisting of Fe hydroxides.

PGM are represented by ferroplatinum and rutheniridosmine associations, as well as by small amount of sperrylite and Pd minerals. More than 65% of grains belong to ferrous platinum and isoferroplatinum; about 30% are minerals of the Ru, Ir and Os system and about 3% are Pd minerals. The degree of grain rounding of ferroplatinum is higher than that of rutheniridosmines and sperrylite, but it is generally low which indicates a short range of their transfer. As impurities, Cu, Rh and Pd are commonly found, Au is found in rare cases.

4 Conclusions

The identified features of native Au and PGM indicate the presence of several types of motherlodes, and possibly, intermediate reservoirs, as well as various distances of placers from sources of supply. The data obtained are compatible with the assumption that input of gold from ledge ores of Au-quartz, Au-quartz-sulfide, and Au-magnetite types widespread in the eastern part of the Kuztetsk Alatau and on the Salair range, as well as from sediments of the Simonovskaya suite. The sources of PGM were rocks of the Ural-Alaskan and ophiolite complexes, fragments of which (in particular the Kaygadatsky massiv) have been established in NWKA.

Acknowledgements. This work was supported financially by the RFBR (Grant 19-05-00464) and government assignment (project No. VIII.72.2.3 (0330-2014-0016)).

Impact Diamonds: Types, Properties and Uses

V. Afanasiev[1(✉)], N. Pokhilenko[1], A. Eliseev[1], S. Gromilov[2], S. Ugapieva[3], and V. Senyut[4]

[1] VS Sobolev Institute of Geology and Mineralogy, Siberian Branch, Russian Academy of Sciences, Novosibirsk, Russia
avp-diamond@mail.ru
[2] Nikolaev Institute of Inorganic Chemistry, Siberian Branch, Russian Academy of Sciences, Novosibirsk, Russia
[3] Diamond and Precious Metal Geology Institute, Siberian Branch, Russian Academy of Sciences, Yakutsk, (Sakha) Yakutia, Russia
[4] Joint Institute of Mechanical Engineering of the NAS of Belarus, Minsk, Belarus

Abstract. Popigai is the world largest crater produced by an impact event. Abundant graphite in the target rocks underwent martensitic transformation into a mixture of high-pressure phases (an aggregate of nanometer cubic diamond and hexagonal lonsdaleite crystals), and some amount of graphite survived as a residual phase. They are of two types: (i) diamonds extracted from tagamites as chips of grains crushed during processing; (ii) yakutites in placers inside and around the crater, which formed at the impact epicenter and dispersed during the event. The impact diamonds possess exceptional abrasive strength, 1.8 to 2.4 times greater than in synthetic diamonds. The outstanding wear resistance, a large specific surface area and a thermal stability (200–250 °C greater than in synthetic diamonds) are favorable for main technological uses. With these properties, impact diamonds are valuable as material for composites and tools.

Keywords: Impact crater · Impact diamond · Yakutite · Abrasion strength

1 Introduction

The Popigai impact crater located in Yakutia (Russia) stores unlimited amounts of impact diamond with exceptional technological properties.

Studies of the Popigai impact crater have a dramatic history. The research was active from 1971, when V.L. Masaitis first proved the impact origin of the crater, to 1986, when the work stopped unexpectedly. Since neither geological surveys, nor engineering testing were undertaken; the testing was impossible because few diamond samples were available. In 2010 we got a collection of impact diamonds (about 3000 carat) from previous work at Popigai and resumed the tests. In this paper we characterized the diamonds as valuable raw for advanced technologies.

2 Popigai Impact Diamonds

A huge crater, about 100 km in diameter, formed 35.7 Ma when a large bolid hit the Earth at the Anabar shield in Siberia (Masaitis et al. 1998). The target Archean gneisses of the Khapchan Group containing abundant crystalline graphite were broken, remolten and partly dispersed outside the crater. Thereby the graphite underwent martensitic transformation into a mixture of high-pressure cubic diamond and hexagonal lonsdaleite; some amount of graphite survived as a residual phase either in aggregates with high-pressure phases or as separate particles. Upon conversion, the graphite reduced in volume by 1.6 and the resulting diamond formed as aggregate of n * 10 – n * 100 nm crystals (Walter et al. 1992). Impact diamonds possess exceptional abrasive strength, greater than synthetic diamonds by 1.8 to 2.4 depending on relative percentages of phases. The ultrahard impact diamonds can be successfully used for composites and tools.

3 Types of Impact Diamonds

The Popigai diamonds comprise yakutites in placers and diamonds hosted by tagamites, the primary impact rocks.

Yakutites formed at the impact epicenter and dispersed during the event. Currently they are found in placers inside and around the crater. The farthest dispersed findings of yakutites occur 550 km away from the crater or even more. The yakutite aggregates of nanometer grains are 0.7–0.8 mm to 1.0 mm in size and have a shapeless morphology or sometimes preserve hexagonal contours of primary graphite. They formed at the highest pressure and consist of cubic diamond and hexagonal lonsdaleite. The presence of graphite appears in Raman spectra but is undetectable by X-ray diffractometry, possibly, because of minor contents.

Tagamite-hosted diamonds are extracted by crushing from very hard host rocks and thus have angular shapes and particle sizes from a few microns to 1 mm. During subsequent flotation, large amounts of fine graphite are extracted along with diamond. These diamonds are likewise aggregates of nanometer diamond, lonsdaleite and graphite grains. Diamond chips look laminated like the precursor graphite. Impact diamonds from tagamite are of two species. A-diamonds are colorless or yellowish, almost fully consisting of the cubic phase; they are the most resistant to wear. Those of B species comprise cubic diamond, lonsdaleite and graphite and are slightly less resistant. Most of impact diamonds, including those we analyzed, are from the Skalnoye deposit in the southwestern flank of the Popigai crater, away from the epicenter. They formed by shock waves while the target rocks were melting and remained enclosed in tagamite, i.e. they originated at lower pressures than yakutites.

Thus, placer yakutites and tagamite-hosted diamonds share the same impact origin but formed at different pressures and evolved in different ways afterwards. The highest-pressure yakutites were dispersed and quenched, and preserved their primary structure, whereas diamonds in tagamites underwent prolonged annealing in tagamite melt. The annealing explains the presence of nitrogen impurity (N3V) which lacks in yakutites.

We explored the Udarnoye and Skalnoye sites within the Popigai crater. The Skalnoye deposit stores up to 100 carat diamond per ton of tagamite, while the total estimated resources exceed 162 billion carat (Masaitis et al. 1998). The resources in the whole crater are actually inexhaustible. The amount of yakutites in placers around the crater may reach 1.5 billion carat (Masaitis et al. 1998).

4 Possible Uses of Impact Diamonds

Possible uses of impact diamonds are based on their exceptional wear resistance. Only tagamite-hosted diamonds were studied. Micropowders were made from a collection of impact diamonds (mixed A and B species) at the Institute of Superhard Materials (Kyiv, Ukraine). The impact diamond micropowder of any grain size shows greater abrasive strength than synthetic diamonds (Table 1).

Table 1. Comparative abrasive strength of impact synthetic diamonds

Impact diamond micropowders		Synthetic diamond micropowders	
Grain size, μm	Abrasive strength, relative units	Grain size, μm	Abrasive strength, relative units
+60	5,05	+60	3,67
60/28	6,53	60/28	3,69
40/20	5,89	40/20	3,54
28/14	5,70	28/14	3,33
14/7	4,85	14/7	2,91
10/5	3,80	10/5	2,16
7/3	2,98	7/3	1,71
5/2	2,20	5/2	1,21

The extremely hard impact diamonds can be superior substitutes for synthetic diamonds in industry. Furthermore, a large specific surface area and relatively high adsorption and thermal stability 200–250 °C greater than in synthetic diamonds, make impact diamonds excellent raw for composites. Sintered composites of "impact diamond - Fe-Ti bond" powdered in a planetary mill to 5 to 50 μm granules show better polishing properties compared with similar powders based on synthetic diamond DSM 20/14 (Table 2).

Table 2. Comparative polishing properties of impact diamonds – Fe-Ti bond and similar powder of synthetic diamonds

Composite powders		Polished material	Weight loss (mg/min)	Stability (min)
Fe-Ti/impact diamond	5/50	Silica	35,4	>30
Fe-Ti/ACM	5/50	Silica	17,8	14

The impact diamond-based abrasive powders used for magnetic-abrasive polishing of silica plates have 1.5–2 times greater strength and 2 times greater stability (lifetime) than their counterparts made from synthetic diamonds (Table 2).

The technological properties of yakutites remain poorly investigated. They obviously have a very high wear resistance, possibly, higher than in tagamitic diamonds, while their grain sizes are suitable for using them in blade tools.

Meanwhile, apart from abrasive properties, there is a large scope of potential uses associated with structure, phase composition, etc. Currently, the only problem is the shortage of diamond specimens for testing. Sufficient amounts of such specimens can be obtained by building a pilot plant for extraction of diamonds.

5 Conclusions

The Popigai impact diamonds possess exceptional abrasion strength, large specific surface area and high thermal stability superior over the respective properties in synthetic diamonds. With these technological properties and actually unlimited resources, the impact diamonds are valuable industrial raw. The development of the Popigai diamond deposit can be successful due to progress.

Acknowledgements. The work was supported by grants 16-05-00873a and 17-17-01154a of the Russian Foundation for Basic Researchers and was carried out as a part of the Project No. 0330-2016-0006.

References

Masaitis VL, Mashchak MS, Raikhlin AI, Selivanovskaya TV, Shafranovskiy GI (eds) (1998) Diamond-bearing impactites from the popigai impact crater. VSEGEI, St. Petersburg

Walter AA, Eryomenko GK, Kvasnitsa VN, Polkanov Y (1992) Carbon minerals produced by impact metamorphism. Naukova Dumka, Kiev

45

Mineralogical Analysis of Glacial Deposits and Titanium Paleoplacers of the East European Part of Russia

N. Vorobyov and A. Shmakova(✉)

Institute of Geology of Komi SC UB RAS, Syktyvkar, Russia
alex.sch92@yandex.ru

Abstract. Mineralogical analysis is one of the main methods to determine sources for paleoplacers and location of source glacial provinces. Our studies defined location of source glacier provinces and sources for titanium paleoplacers of the East European Part of Russia.

Keywords: Minerals · Reconstruction · Glacial deposits · Titanium paleoplacers

1 Introduction

Mineralogical analysis is the most significant method in reconstructing sources paleoplacers and glacial deposits (boulder loams), determining location of source for glacial provinces. The paper presents two objects - glacial deposits of polar and vychegda horizons, Middle Devonian and Middle Triassic titanium paleoplacers.

2 Results and Discussion

Quaternary deposits have been studied in two areas of Pechora lowland: in the northeast in the basin of the Padymeytyvys river and northwest in the basin of the Kui river. Titanium paleoplacers were studied at the Kydzarasyu river (Preural Foredeep) and Middle Timan (Pizhma paleoplacer).

Relations between mineral composition of Quaternary deposits and underlying bedrocks are very important. During active exaration activity of the glacier, the underlying rocks controlled composition of boulder loams. The formation of heteroaged horizons of boulder loams is associated with the Northeastern (Paykhoy-Ural-Novaya Zemlya) and Northwestern (Fennoscandinavia) terrigenous-mineralogical source provinces. The rocks of the eastern province are characterized by higher levels of epidote and ilmenite fraction, and the northwestern ones - amphiboles and garnets.

In the basin of the Padymeytyvys river two horizons of boulder loams were drilled. In the lower (Vychegodsky) horizon, epidote (19.9%) and siderite (21.2%) dominate in the heavy fraction. In the composition of the heavy fraction of the upper (polar) horizon, pyrite (22.1%) and epidote (20.7%) have maximum concentrations, siderite content (11.7%) and ilmenite (14.7%) are high. Pyrite and siderite indicates the relation between the glacier and Triassic and Permian underlying rocks, and ilmenite may be associated with Uralian rocks.

In the valley of the Kui river one horizon of boulder loam – polar – was drilled. The heavy fraction is represented by amphibole (13.6%) - garnet (18%) - epidote (20.8%) associations with the increased content of pyrite (7.7%). High concentrations of amphiboles (13.6%) and garnets (18%) may indicate the relation between the glacier and the rocks of Fennoscandinavia.

The number of amphiboles in boulder loams decreases from west to east and amounts to the first percent. High total contents of pyrite and siderite in the north-east of Pechora lowland are typical of local Mesozoic source province.

Heavy fraction of the paleoplacer at the Kydrasyu river is represented by ilmenite (45.19%), epidote (23.21%), magnetite (15.76%), amphibole (6.69%), chrome spinelide (5.65%), garnets (2.06%). Also zircon, leucoxen, rutile, hematite, kyanite, martite, pumpellyite, staurolite are present (0.04–0.4%). The high content of ilmenite indicates the relation between the paleoplacer and igneous rocks. Epidote, garnet, kyanite, staurolite, and pumpellyite found in the heavy fraction are of metamorphic origin. The morphology of zircons is characteristic of minerals formed in igneous rocks of medium composition. The peculiarity of these minerals is their good preservation. This may be due to the nearby source. The mineral composition indicates that the source of minerals was not far from its burial. Titanium minerals and satellite minerals were most likely of magmatic and metamorphic origin.

The mineral composition of titanium placers often contains stable minerals and there are no unstable ones, although both are present in the bedrocks. The main reason is that the paleoplacer formed due to erosion of the weathering crust. An example of a paleoplacer formed that way can be Pizhma paleoplacer (Middle Timan) formed due to erosion and redeposition of weathering crusts on Riphean shales. It's mineral composition in contrast to the paleoplacer at the Kydzarasyu river, is very poor. The main part is represented by leucoxenized ilmenite and leucoxene. Tourmaline, garnet and zircon are found in single units.

3 Conclusions

We determined location of the source glacier provinces and sources for the titanium paleoplacers of the East European part of Russia by the mineralogical analysis.

46

Gold-Silver Natural Alloy of Chromitites from the Kamenushinsky Massif

A. Minibaev(✉)

Mining University, Saint-Petersburg, Russia
a.m.minibaev@yandex.ru

Abstract. A detailed study of the chromitites clinopyroxenite Kamenushinsky massif in the middle Urals has allowed to allocate from them the gold-silver alloy. The study of the chemical composition, morphology of gold-silver alloys, peculiarities of their placement allowed not only to determine their genetic relationship with platinum-bearing chromites, but also to make an assumption about the formation of chromitites as a result of a single geological process.

Keywords: Urals Platinum Belt · Kamenushinskiy massif · Chromite-platinum mineralization · Gold

1 Introduction

The history of the development of platinum placers of the Urals dates back nearly a two centuries and a distinctive feature is the permanent presence of gold and its natural alloys along with the extraction of the platinum groups minerals (hereafter - PGM) (Vysotsky 1913). It is known that the main source of the PGM is a process of a zonal massifs erosion of the Ural-Alyaskan type. Nevertheless, the question of a primary source of gold in placers is remained as a controversial issue. Thus, for example, in research works of N. Vysotsky (Vysotsky 1913) was given a support of the idea that the appearance of gold in platinum placers was due to the weathering of an acid rocks, but its absence in some zonal massifs of the Urals Platinum Belt (hereafter - UPB) keeps the matter in abeyance. In the 70s of the 20th century during setting deposit into exploitation, located within the Kachkanar zonal massif UPB, for the first time gold was determined and described from the bed rock (Fominykh et al. 1970). In spite of this fact, further the question of the nature of the appearance of gold in the ores of the UPB massifs was not under active discussion.

Inside the Nysyamsky platinum placer field – one of the main leaders of the platinum mining of the Urals – also the presence of gold was reported (Vysotsky 1913). The source of the placer formations was Kamenushinsky dunite-clinopyroxenite massif which had a potential for of bed chromite-platinum mineralization (Ivanov 1997; Tolstykh et al. 2011, Minibaev et al. 2015). During the study platinum-bearing chromitites were divided into two petrographic types with characteristic features: vein-imbedded and massive (Minibaev 2018). In addition to study of PGM features from the chromitites of both petrographic types, the scientific observation of the determined Au-Ag alloys is under a great interest

2 Methods and Approaches

A study of morphology and chemical composition of Au-Ag natural alloys and its bearing chromitites was carried out by the scanning electron microscope Carl Zeiss EVO (OPTEC, Moscow), equipped with attachment: EDS (energy dispersive X-ray spectrometer) and BSD (Backscattered Electron Detector).

3 Results and Discussion

Au-Ag natural alloys was determined as impurities of irregular crystallographic habitus in chromespinelides of vein-imbedded (Fig. 1a) and massive (Fig. 1b) chromitites.

The size of the aggregates is around 4–6 microns. The chemical composition of Au-Ag alloys (Table 1) corresponds to the compositions of similar objects founded in the ore chromospinelides of the Konder massif (Pushkarev et al. 2015).

The similar composition of Au-Ag alloys, its presence directly in chromespinelides, also the absence of traces of visible deformations of the mentioned indicates that its formation does not correspond to the processes of serpentinization or overlaid hydrothermal processes (as it was known before), it corresponds to the fact that inclusions were gained into chromespinelides during crystallization. Inclusions such as Cr and Fe in Au-Ag alloys are the evidence of inherited condition of mineralization and chemistry features of the ore-forming system.

Table 1. Chemical composition of Au-Ag alloys from the chromitites of the Kamenushinsky massif (mass. %)

№ point	Au	Ag	Cr	Fe	Sum
1	81,45	10,84	3,84	2,02	98,15
2	81,21	10,47	3,55	1,89	97,12

Notes: 1 - from vein-imbedded chromitites; 2 - from massive chromitites.

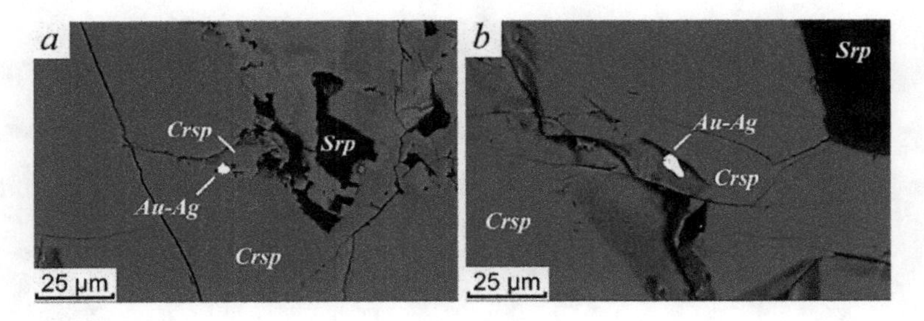

Fig. 1. Determination of grains Au-Ag natural alloys in chromitites: a - vein-imbedded; b - massiv

4 Conclusions

The obtained results are correlated with the conclusions about the syngenetic nature of the platinum-bearing vein-imbedded and massive chromitites, where the latter are characterized by a later origin relating to the final stage of evolution of the ore-forming system (Minibaev 2018). Also these results can be used not only to confirm the hypothesis about the common substance source of two petrographic types of chromitites, but also allow to make the conclusion that the formation of chromitites is generally resulted from a single geological process.

Acknowledgements. Work was done with the support of government program № 5.12856.2018/8.9.

References

Fominykh VG, Uskov ED, Volchenko YuA (1970) Gold in the ores of the Kachkanar massif in the Middle Urals. In: Questions of the Geology of the Gold Deposit, Tomsk, pp 31–36

Ivanov OK (1997) Concentric-zonal pyroxenite-dunite massifs of the Urals. Ed. Ural University, Ekaterinburg, 488 p

Minibaev AM (2018) About the origin of chromite-platinum mineralized zones of the clinopyroxenite-dunite Kamenushinsky massif of the middle Urals. In: Materials of the Eighth Russian Youth School with International Participation "New in the Knowledge of the Processes of Ore Formation". IGEM RAS, Moscow, pp 255–258

Minibaev AM, Stepanov SYu (2015) Perspectives of the identification of chromite-platinum mineralization in the rocks of the Kamenushinsky zonal clinopyroxenite-dunite massif (Middle Ural). In: Proceedings of the Fifth Russian Youth School with International Participation "New in the Knowledge of the Processes of Ore Formation". IGEM RAS, Moscow, pp 145–148

Pushkarev EV, Kamenetsky VS, Morozova AV, Hiller VV, Glavatskikh SP, Rodemann T (2015) Ontogeny of ore chromespinelide and composition of inclusions as an indicators of pneumatolyte-hydrothermal formation of platinum-bearing chromite of the Konder massif (Aldan shield). Geol Ore Deposits 57(5):394–423

Tolstykh ND, Telegin Y, Kozlov AP (2011) The native platinum of the Svetloborsky and Kamenushinsky massifs of the Ural platinum-bearing belt. Geol Geophys 52(6):775–793

Vysotsky NK (1913) Platinum deposits of Isovsky and Nizhne-Tagilsky regions of the Urals. In: Proceedings of the Geological Committee. New edit, no. 62, 692 p

47

Biooxidation of Copper Sulfide Minerals

Yu. Elkina[1,2(✉)], E. Melnikova[2], V. Melamud[2], and A. Bulaev[1,2]

[1] Faculty of Biology, Lomonosov Moscow State University, Moscow, Russia
yollkina@mail.ru

[2] Research Center of Biotechnology RAS, Moscow, Russia

Abstract. The effects of temperature and the presence of NaCl on bioleaching of chalcopyrite, enargite, and tennantite were studied. Rate of copper extraction from all minerals depended on temperature and was the highest at 45–50 °C. NaCl addition increased rate of copper extraction from chalcopyrite but led to the decrease in copper extraction from enargite and tennantite.

Keywords: Bioleaching · Chalcopyrite · Enargite · Tennantite · Acidophilic microorganisms

1 Introduction

Copper and zinc are mainly extracted from sulfide ores using pyrometallurgical techniques. Pyrometallurgical processing of arsenic containing ores is a problem due to the emission of toxic gases (Filippou et al. 2007). Biohydrometallurgy is widely used to process gold bearing concentrates, and may also be used to extract non-ferrous metals arsenic-containing concentrates (Neale et al. 2017). The goal of the present work was to study copper bioleaching from arsenic-containing minerals and chalcopyrite at different temperatures and in NaCl presence.

2 Methods and Approaches

Chalcopyrite ($CuFeS_2$), enargite (Cu_3AsS_4), and tennantite ($Cu_{12}As_4S_{13}$) as well as mixed culture of acidophilic microorganisms oxidizing ferrous iron and sulfur compounds were subjects of the study. The experiments were carried out in flasks with 100 ml of nutrient medium supplemented and 2 g of milled minerals (P_{100} 75 μM) on a rotary shaker at temperatures from 40 °C to 60 °C for 30 days. In one variant of the experiment, nutrient medium was supplemented with100 mm NaCl.

3 Results and Discussion

The results of the experiments (rates of copper extraction) are shown in Table 1.

Table 1. Rate of copper extraction from the minerals for 30 days (%)

Variant of the experiment	Chalcopyrite ($CuFeS_2$)	Enargite (Cu_3AsS_4)	Tennantite ($Cu_{12}As_4S_{13}$)
40 °C	14.33 ± 0.08	8.18 ± 0.65	15.33 ± 0.24
45 °C	17.84 ± 1.69	12.89 ± 1.79	26.16 ± 1.33
50 °C	25.29 ± 4.15	14.04 ± 0.95	18.43 ± 0.84
50 °C, 100 NaCl	33.25 ± 0.12	5.91 ± 1.09	13.04 ± 0.03
55 °C	26.75 ± 1.57	14.39 ± 0.01	14.84 ± 0.01
60 °C	17.32 ± 1.01	5.86 ± 2.18	12.83 ± 0.17

We showed that the rate of copper extraction from all minerals depended on temperature and was low at 40 °C. Extraction rate at 60 °C was also low as this temperature might inhibit microbial activity. Addition of NaCl increased rate of copper extraction from chalcopyrite that was a well known phenomenon. At the same time, NaCl addition led to the decrease in copper extraction rate.

4 Conclusions

The results obtained suggest that the efficiency of copper sulfide minerals depended on temperature, while NaCl addition did not allowed increasing the rate of copper bioleaching from arsenic-containing minerals in contrast to chalcopyrite. This fact should be taken into consideration when planning laboratory scale trials on bioleaching of copper sulfide concentrates.

Acknowledgements. The work was supported by the President Grant of the Russian Federation, grant No. MK-6639.2018.8.

References

Filippou D, St-Germain P, Grammatikopoulos T (2007) Recovery of metal values from copper – arsenic minerals and other related resources. Miner Process Extr Metall Rev 28(4):247–298

Neale J, Seppälä J, Laukka A, van Aswegen P, Barnett S, Gericke M (2017) The MONDO minerals nickel sulfide bioleach project: from test work to early plant operation. Solid State Phenom 262:28–32

Microbial Processes in Ore-Bearing Laterite at the Tomtor Nb-REE Deposit: Evidence from Carbon Isotope Composition in Carbonates

V. Ponomarchuk(✉), E. Lazareva, S. Zhmodik, N. Karmanov, and A. Piryaev

Institute of Geology and Mineralogy SB RAS, Novosibirsk, Russia
ponomar@igm.nsc.ru

Abstract. The unique Nb-REE deposit is located within the Tomtor complex of ultramafic alkaline and carbonatite rocks in the northern Sakha Republic (Yakutia) (Kravchenko and Pokrovsky 1995; Dobretsov and Pokhilenko 2010; Lazareva et al. 2015). Ores reside in three layers (Severny, Yuzhny, and Buranny sites) which fill depressions in subsided profiles of weathered carbonatites. Judging by stable isotope analysis, carbonates from laterite weathering profiles at the Tomtor Nb-REE-deposit formed by different mechanisms, including microbially mediated organic-clastic sulfate reduction and anaerobic oxidation of methane.

Keywords: Tomtor · Nb-REE-deposit · Laterite · Carbonates · C-isotopes · Microbial · Methane

1 Introduction

Rare earth elements have been broadly used in advanced technologies. They often occur in carbonatite deposits among which laterite profiles of weathered carbonatites are most attractive commercial targets. A unique Nb-REE deposit is located within the Tomtor complex of ultramafic alkaline and carbonatite rocks in the northern Sakha Republic (Yakutia) (Kravchenko and Pokrovsky 1995; Dobretsov and Pokhilenko 2010; Lazareva et al. 2015). Ores reside in three layers (Severny, Yuzhny, and Buranny sites) which fill depressions in subsided profiles of weathered carbonatites. Lately, evidence has been obtained that localization of elements in ore-bearing beds may have biotic controls (Lazareva et al. 2015). The possible role of microorganisms in ore formation has not been discussed yet in the literature on laterite profiles from carbonatite deposits, including the giants of Mountain Weld (Australia) or Araxa and Catalão (Brazil). In this respect, the Tomtor deposit is remarkable by a contribution of biogenic and bacterial processes to its formation, besides magmatism and high-temperature hydrothermalism that are common to all carbonatite deposits. Biogenic agents are usually identified from the carbon isotope composition of rocks controlled by their interaction with fluids. The discussed stable isotope compositions of carbonates from ore-bearing laterites at the Tomtor deposit provide evidence of possible microbial mediation.

2 Methods and Approaches

The stable isotope compositions were analyzed in three samples (TM-592, TM-590, and BH 101) from laterite ore zones with ~1%, 0.6%, and 1.6% REE, respectively. Extraction of monofractions for analysis is difficult because the laterite samples are fine grained while the carbonates are zoned. The problem can be solved by using selective acid extraction based on progressively slower reactions of carbonates with H_3PO_4 in the stoichiometric series: calcite → dolomite → ankerite → siderite → rhodochrosite. According to experimental evidence (Al-Aasm et al. 1990), the reaction duration sufficient for $\delta^{13}C$ determination is 1 h for calcite, 24 h for dolomite, and 5–7 days for siderite and rhodochrosite. The C and O isotope compositions of carbonates are determined by digestion in anhydrous H_3PO_4 followed by measurements on a FINNIGAN MAT-253 mass spectrometer with a GasBench II analyzer, in a stream of pure helium. The GasBench unit was also used for preconditioning of samples digested at 70 °C. The results are quoted in (‰) relative to the Vienna Peedee belemnite PDB standard for carbon and relative to SMOW for oxygen. The analytical errors were ±0.1‰ for carbon and 0.15‰ for oxygen. The composition and morphology of minerals were studied at the Analytical Center of IGM (Novosibirsk) on a Tescan MIRA 3 LMU scanning electron microscope with Oxford Instruments Nanoanalysis Aztec Energy/INCA Energy 450+ XMax 80 and INCA Wave 500 analyzers, applicable to scan nanometer particles.

3 Results and Discussion

Sample TM-590 is composed mainly of goetite, siderite (with a Mn impurity), calcite, and rhodochrosite. Goethite occurs either as dripstone, with concentric zonation, or as pseudomorphs after a disappeared mineral. The carbonates exist as zoned anhedral grains, with calcite being the latest phase. Fine apatite crystals, pyroxene, monazite, pyrochlore, and sphalerite are present in lesser amounts. The rock-forming phases in TM-592 are goethite, siderite, and calcite and the accessories are apatite and monazite. Apatite occurs as fine prismatic euhedral grains in calcite. Sample BH 101 consists of finely intergrown Fe-chlorite (chamosite) and apatite. The apatite-chamosite aggregate encloses clearly seen large rhodochrosite grains with siderite in their core. The two phases have a distinct boundary but have inherited orientations of the crystallographic axes. The accessory phases are TiO_2, submicrometer monazite platelets, sphalerite, and galena.

Synthesis of C and O isotope data from magmatic and postmagmatic (hydrothermal, metasomatic) carbonates in different carbonatite deposits world wide shows a large range of values from −10 to +3‰ (PDB) for $\delta^{13}C$ and from +6 to +30‰ (SMOW) for $\delta^{18}O$ (Deines 1989).

The $\delta^{13}C$ values obtained for the Tomtor samples (Table 1) are lower. The $\delta^{13}C$ patterns in carbonates are controlled by the isotope composition of bicarbonate. Low $\delta^{13}C$ in bicarbonate is due to oxidation of organic matter where $\delta^{13}C$ can reach −32‰ (Oleary 1988). In the course of microbially mediated sulfate reduction of geopolymers and biopolymers in aerobic conditions, the $\delta^{13}C$ values of bicarbonate that forms by reaction (1) and those of carbonate produced later by reaction (2) are inherited from the precursor component.

$$2CH_2O + SO_4^{2-} \rightarrow 2HCO^{3-} + H_2S \quad (1)$$

$$2HCO^{3-} + Me^{2+} = MeCO_3 + CO_2 + H_2O \quad (2)$$

There are two facts that implicitly support the above considerations: laterite contains organic remnants (Lazareva et al. 2015) that are electron donors in reaction (1); the samples contain sulfides, including framboidal pyrite, which formed at the account of H_2S released in reaction (1). However, this model can account only for the isotope composition of the calcite component in TM-590, while other $\delta^{13}C$ values (−24.8‰ in TM-590) are below −32‰, and may have formed with participation of methane. Biogenic methane has −55‰ to −80‰ $\delta^{13}C$, and the $\delta^{13}C$ values in bicarbonate resulting from microbially mediated methane oxidation (react/3) and in carbonate that formed later by reaction (2) will be low.

$$CH_4 + SO_4^{2-} \rightarrow HCO^{3-} + HS^- + H_2O \quad (3)$$

Table 1. $\delta^{13}C$ and $\delta^{18}O$ (‰) in different fractions of carbonates

Sample		1 h	24 h	7 days
TM-590	$\delta^{13}C$	−29.6	−39.2	−24.8
	$\delta^{18}O$	+12.0	+16.1	+14.6
TM-592	$\delta^{13}C$	−31.2	−44.1	−37.1
	$\delta^{18}O$	+7.6	+13.6	+15.1
BH 101	$\delta^{13}C$	−59.0	−54.6	–
	$\delta^{18}O$	+9.4	10.5	–

Note that a low value of $\delta^{13}C$ was previously reported (Pokrovsky 1990; Kravchenko and Pokrovsky 1995) at the Tomtor deposit, but the mineralogical and geochemical characterization of the sample is not given.

4 Conclusions

The reported isotope data show that organic clastic sulfate reduction and anaerobic methane oxidation with participation of microbial communities were among key mechanisms responsible for the formation of carbonates in laterite profiles of the Tomtor Nb-REE deposit.

Acknowledgements. The study was supported by grant 18-17-00120 from the Russian Science Foundation.

References

Al-Aasm IS, Taylor BE, South B (1990) Stable isotope analysis of multiple carbonate samples using selective acid extraction. Chem Geol Isot Geosci Sect 80(2):119–125

de Toledo MCM, de Oliveira SMB, Fontan F, Ferrari VC, de Parseval P (2004) Mineralogia, morfologia e cristaloquímica da monazita de Catalão I (GO, Brasil). Braz J Geol 34(1):135–146

Deines P (1989) Stable isotope variations in carbonatites. In: Bell K (ed) Carbonatites, Genesis and Evolution. Unwin Hyman, London, pp 301–359

Dobretsov NL, Pokhilenko NP (2010) Mineral resources and development in the Russian Arctic. Russ Geol Geophys 51(1):98–111

Kravchenko SM, Pokrovsky BG (1995) The Tomtor alkaline ultrabasic massif and related REE-Nb deposits, Northern Siberia. Econ Geol 90(3):676–689

Lazareva EV, Zhmodik SM, Dobretsov NL et al (2015) Main minerals of abnormally high-grade ores of the Tomtor deposit (Arctic Siberia). Rus Geol Geophys 56(6):844–873

Olary MO (1988) Carbon isotopes in photosynthesis. Bioscience 38(5):328–336

Pokrovsky BG, Belyakov AYu, Kravchenko SM et al (1990) Origin of carbonatites and ore-bearing rocks of the Tomtor massif, NW Yakutia, according to isotopic data. Geokhimiya (9):1320–1329. (in Russia)

Ribeiro CC, Brod JA, Junqueira-Brod TC, Gaspar JC, Petrinovic IA (2005) Mineralogical and field aspects of magma fragmentation deposits in a carbonate–phosphate magma chamber: evidence from the Catalão I complex, Brazil. J South Am Earth Sci 18(3–4):355–369

Authentic Semi-Precious and Precious Gemstones of Turkey: Special Emphasis on the Ones Preferred for Prayer Beads

E. Çiftçi[1(✉)], H. Selim[2], and H. Sendir[3]

[1] Department of Geological Engineering, Faculty of Mines, ITU, 33469 Maslak, Istanbul, Turkey
eciftci@itu.edu.tr

[2] Faculty of Engineering, Department of Jewellery Engineering, ITICU, 34840 Küçükyalı, Istanbul, Turkey

[3] Faculty of Engineering and Architecture, Department of Geological Engineering, EOGU, 26480 Odunpazarı, Eskişehir, Turkey

Abstract. There are 6 semi-precious gemstones in Turkey that are the most significant in terms of abundance and authenticity. These include smoky quartz, blue chalcedony, chrysoprase (aka Şenkaya emerald), diaspore (aka sultanite/zultanite), sepiolite (aka meerschaum/Eskişehirstone), and jet (aka Oltustone). The smoky quartz occurs in the south of Büyük Menderes Basin within metamorphic rocks of the massif. Chalcedony occurrences are in Çanakkale and Sarıcakaya (Eskişehir). Chrysoprase is acquired in Bursa, Alaşehir (Manisa) as yellowish green in color, Biga (Çanakkale), Sivrihisar (Eskişehir) and Şenkaya (Erzurum) as dark green in color. Diaspore occurs in Milas (Muğla), Söke (Aydın), Tire (İzmir), Bolkardağı Gerdekkilise area and Saimbeyli (Adana). Sepiolite occurrences are limited to Kıbrısçık-Köroğlu (Bolu), Eskişehir and Konya. Jet (aka Oltustone) occurs in Oltu area (Erzurum). These are the major varieties that has been used to produce both jewelry and ornamental objects among which the prayer beads are one of the most indispensable object of oriental cultures in particular. These are also considered to be the authentic gems of Turkey and exported as raw and processed in into many forms. As for the prayer beads, lightness, color, durability, hardness, and cost are the main criteria. Thus, the Oltustone has been the major source for decades. However, the Eskişehir stone and blue chalcedony are becoming popular as well.

Keywords: Turkish gems · Oltustone · Eskişehirstone · Diaspore · Smoky quartz · Blue chalcedony

1 Introduction

Today, the jewelry sector is living its golden age. With increasing technical capacity and manual aptitude, gold, silver, platinum etc. metals can be embroidered with the precious and semi-precious gems and delivered to consumers. This sector is a fashion-driven dynamic sector and subjected to rapid changes and innovations that are controlled and dictated by the fashion trends and demands of the consumers. Companies are forced to keep up with those and both develop and manufacture the products accordingly. Use of the semi-precious and precious gems is likewise chosen according to the demand by the fashion and modern trends. Consumers religiously follow those two facts. Some of those are considered to be semi-precious gems of Turkey including smoky quartz, chalcedony, chrysoprase, diaspore, sepiolite, and jet (Selim 2015).

2 Methods and Approaches

Representative samples that acquired from the deposits were prepared for further analytical studies for comprehensive characterization. Polished sections and powdered samples were prepared for Transmitted and Reflected Light Microscopy integrated with Optical Cathodoluminescence, Electron Probe Microanalysis (EPMA) and Secondary Electron Microscopy-Energy Dispersive Spectroscopy (SEM-EDS), X-ray Diffraction (XRD), X-ray Fluorescence (XRF) and Inductively-Coupled Plasma Mass Spectroscopy analyses.

3 Results and Discussion

3.1 Smoky Quartz

Smoky quartz is one of the varieties of quartz with dark gray-black color (Fig. 1). It occurs rarely in the Alps, Colorado and California (North America). Its occurrence in Turkey is limited to the Menderes Massif, one of the metamorphic complexes of Turkey. It occurs in metamorphic rocks composed of mafic minerals that expose in southern portion of the massif in Great Menderes basin (İlhan 2012). Location is the city of Aydın and covers a few rustic villages within Çine, Ovacık and Koçarlı county limits. Host-rocks are variable degree metamorphic rocks including slate, phyllite, schist, and gneiss. Smoky quartz occurs locally within the fractures of these rocks. Its major use is as jewelry and ornamental objects. Type locality is Mersinbelen village in the area.

Fig. 1. Macroscopic views of natural smoky quartz and blue chalcedony (Selim 2015)

3.2 Chalcedony

Chalcedony occurs in a few locations including Bolu (Kıbrısçık), Ankara (Beypazarı), Afyon (Bayat), İzmir (Aliağa, Bergama, Seferihisar-Yukarıdoğanbey), Tokat (Zile) in lilac-purple colors, Yozgat (Çekerek) with amethyst as botryoidal masses, Çanakkale (center) and Eskişehir (Sarıcakaya) in blue color. Druzy chalcedony occurs in Sivas (Yıldızeli) (İlhan 2012). Chalcedony is a silica variety mineral. Its color may vary from deep blue to faint blue, purple in places, translucent and opaque (Fig. 1). Among these, blue chalcedony occurrence in Turkey is located in Mayıslar village (Sarıcakaya-Eskişehir) owned and exploited by Kalsedon Co. (Hatipoğlu et al. 2010a). Blueish color is either due to the water content or abnormally high barium content. This is also considered to be the type locality. Its major use is as jewelry and ornamental objects.

3.3 Chrysoprase

Chrysoprase or chrysoprasus is a gemstone variety of chalcedony that contains small quantities of nickel. The color varies from faint light green to apple green (Fig. 2). It is translucent and opaque. It worldwide occurrences are reported from USA (California, Oregon, and Arizona) and Eastern India. In Turkey, it occurs in Alaşehir (Manisa) in yellowish green, Dikmendağı area in Biga (Çanakkale) deep green, Sivrihisar (Eskişehir) in deep green, İkizce village (Bursa), Zümrüt and Turnalı villages of Şenkaya county (Erzurum). In Erzurum, due to it locality and most probably to the color, it is erroneously named as *"Şenkaya Emerald"*. It has naturally formed patterns and quasi figures that make small objects even more fascinating.

Fig. 2. Macroscopic views of natural chrysoprase and diaspore (Selim 2015)

3.4 Diaspore

Diaspore is one of the minerals forming the bauxite ore. It is the least hydrous Al-oxyhydroxide mineral in the bauxites. Diaspore becomes the main, even only mineral of bauxites when the bauxite ore gets metamorphosed. Such ores are better called as diasporite, since those are monomineralic. This is very common occurrence especially in the Menderes Massif (western Anatolia). Diaspore crystals sometimes grow into really gigantic sizes. Then becomes gem material. Diaspore naturally occur in yellowish green and light green colors (Fig. 2). In addition to the metabauxites, diaspore occurrences are also reported in laterites and Al-rich clays, in recrystallized limestones, with corundum, in some alkali pegmatites as late stage hydrothermal mineral, and in magmatic rocks as a result of recrystallization of Al-rich xenoliths (Hatipoğlu et al. 2010b; Kumbasar and Aykol 1993). However, they hardly grow into gem size and quality. In Turkey, type locality is Milas county (Muğla) (Lake Bafa vicinity and area in between Milas-Yatağan counties). There are reports on its existence in Söke (Aydın), Tire (İzmir), in central Anatolia (Gerdekkilise-Bolkardağ) and Saimbeyli county (Adana), however, these are not at mineable scale. Gem quality diaspores are mostly used in jewelry. Two names were adopted, also trademarks, for such diaspores, *"Sultanite or Zultanite"* in honor of Ottoman Sultans and *"Csarite"*. It has very unique phenomenal color changing property and rarity thus becoming more popular. It is considered as "diamond of the future" by many.

3.5 Sepiolite

Sepiolite, also known as meerschaum, is a soft white clay mineral - a complex magnesium silicate, a typical chemical formula ($Mg_4Si_6O_{15}(OH)_2{\cdot}6H_2O$). It can be present in fibrous, fine-particulate, and solid forms. Sepiolite is opaque and off-white, grey or cream color, breaking with a conchoidal or fine earthy fracture, and occasionally fibrous in texture (Fig. 3). There many reports on its occurrence worldwide. It is generally associated with carbonate/evaporate sequences, sedimentary in origin. However, it also occurs in Kıbrısçık (Bolu) as a result of hydrothermal alteration of glassy tuffs of Middle Miocene Deveören volcanics in Köroğlu (Galatya) volcanic belt (Irkeç and Ünlü 1993). Although there are occurrences worldwide, most of the sepiolite of commerce is obtained chiefly from Sepetçi, Margı, Sarısu, Kayı, Gökçeoğlu, İmişehir, and Türkmentokat villages of the city of Eskişehir in Turkey. Nemli (Kütahya) is also important for its alike occurrences. It is mostly used to make pipes and small objects, paring beads. Calcite, gypsum and dolomite often company sepiolite.

Fig. 3. Macroscopic views of worked and raw Eskişehirstone and raw Oltustone (Selim 2015)

3.6 Jet/Oltu Stone

Jet (literal name) or Oltustone (locality name), is an opaque black coalified fossilized drift wood of trees of the family Araucariaceae which is 180 million years old (Fig. 3).

It is formed from the high-pressure decomposition of wood. Jet is chemically related to brown coal, or lignite, but it is more solid and tough, looks more like obsidian than a coal (Toprak 2013). Jet has been used both as a talisman and a jewel for over four thousand years. In Turkey, it occurs and mined in Oltu county (Erzurum) where more than 300 locations around Tutlu (Lispek), Yeşilbağlar (Norpet), Güllüce, and Güzelsu villages. Jet is also found in other locations around the world. However, only the British and Spanish deposits in addition to the Turkish occurrences have been worked commercially. Jet is called *"Oltustone"* in Turkey due to its type locality (Ciftci et al. 2004). It is mostly used in jewelry and to make praying beads. Framboidal pyrites and clay pockets commonly occur in jet.

4 Conclusions

Among the many, 6 semi-precious stones are the most common and renowned: smoky quartz, Eskişehirstone, Oltustone, sultanite, blue chalcedony, and Şenkaya emerald. Major use of these include jewelry (sultanite, Oltustone, smoky quartz, blue chalcedony, and Şenkaya emerald), ornamental objects (Eskişehirstone, Oltustone, smoky quartz, blue chalcedony, and Şenkaya emrald), prayer beads (Oltustone, blue chalcedony, and Eskişehirstone). Sultanite (diamond of the future) is a rising star in the precious stones due to its only occurrence in Milas area (Muğla) and unique properties.

References

Ciftci E, Yalcin MG, Yalcınalp B, Kolaylı H (2004) Mineralogical and physical characterization of the Oltustone, a Gemstone Occurring around Oltu (Erzurum-Eastern Turkey). In: 8th ICAM, São Paulo-Brazil, pp 537–538

Hatipoğlu M, Babalık H, Chamberlain SC (2010a) Gemstone deposits in turkey. Rocks Miner 85 (2):124–133

Hatipoğlu M, Türk N, Chamberlain SC, Akgün AM (2010b) Gem-quality transparent diaspore (Zultanite) in bauxite deposits of the Ilbir Mountains, Menderes Massif. SW Turk Miner Deposita 45(2):201–205

İlhan NN (2012) Türkiye'nin Mücevher Taşları Haritası. I. Türkiye Mücevher Taşları Sempozyumu, Bildiriler Kitabı, İstanbul, pp 33–38

İrkeç T, Ünlü T (1993) Volkanik kuşaklarda hidrotermal Sepiyolit oluşumuna bir örnek: Kıbrısçık (Bolu) Sepiyoliti. MTA Dergisi, Ankara, vol 115, pp 99–118

Kumbasar I, Aykol A (1993) Mineraloji. İstanbul Teknik Üniversitesi Kütüphanesi, Sayı: 1519, İstanbul

Selim HH (2015) Türkiye'nin değerli ve yarı değerli mücevher taşları. İstanbul Ticaret Odası (İTO) Yayınları No: 2014/4, 102s, İstanbul

Toprak S (2013) Petrographical properties of a semi-precious coaly stone, Oltu stone, from eastern Turkey. Int J Coal Geol 120:95–101

50

Noble Metal Mineralization of the PGM Zone "C" of the East-Pana Layered Intrusion

O. Kazanov[1], G. Logovskaya[2(✉)], and S. Korneev[2]

[1] Moscow Branch, FSUE "All-Russian Scientific-Research Institute of Mineral Resources named after N.M. Fedorovsky", Moscow, Russia
[2] Institute of Earth Science, Saint-Petersburg State University, Saint-Petersburg, Russia
galkanuu@gmail.com

Abstract. The first study of the noble metal mineralization of the ore zone "C" of the East-Pana massif allowed to divide it into two types: early magmatic low-sulfide and late post-magmatic proper PGE associations.

Keywords: Fedorovo-Pana massif · East-Pana massif · Stratiform mineralization · Platinum metal mineralization · Early magmatic and late post-magmatic platinum mineral associations

1 Introduction

Since the late 1980s in Kola Peninsula the platinum content, associated with ultramafic-mafic magmatic complexes, has been studied. The Fedorovo-Pana early proterozoic layered intrusion of the peridotite-pyroxenite-gabbronorite formation has been recognized as the most promising object for discovering the industrial reserves of complex low sulfide PGE ores.

2 Methods and Approaches

The following research methods were used in the work: mineralogical description of sections of boreholes, study of samples of rocks, slides and polished sections, microprobe analysis of transparent polished thin sections from ore intervals, statistical treatment of chemical analysis data with program Statistica v.6.1 and then analysis and interpretation of results. These methods allowed to identify similarities and differences in the ore intervals and their mineralization, and to confirm the assumption of a various genesis of PGE mineralization in the ore zone "C" (Subbotin et al. 2012).

3 Results and Discussion

Based on the results of the study of ore mineralization within the ore zone "C", two associations of noble metal minerals were identified. The first, early, magmatic association is represented by platinum and palladium sulfides (Fig. 1), native gold and silver, moncheite, kotulskite and ferroplatinum, forming inclusions in the magmatic major sulphide and rock-forming silicate minerals. There is a significant positive correlation dependence of Pt, Pd, Au with Cu, Ni and S. It is indicated in the Sungiyok area. Later, postmagmatic association is represented by sperrylite (Fig. 1), mertieite-I, mertieite-II, temagamite, telargpalite, stibiopalladinite and kotulskite, associated with secondary silicate and later sulphide (millerite) minerals. They develop along cracks and veins in the rock-forming and early sulfide minerals. For the second type, correlations with Cu are not characteristic, with S there is a significant negative correlation. This association is observed in the Chuarvy area. Both types of mineralization are characterized by abnormally low values of Pd/Pt = 0.1–1.7 (Voytekhovich et al. 2008).

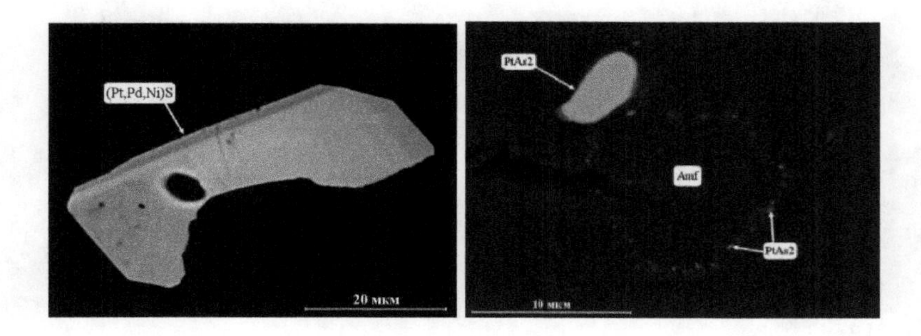

Fig. 1. Subhedral zonal aggregate of braggite ((Pt, Pd, Ni)S - early magmatic association of PGE mineralization (left); Spot edge around amphibole and micron streak of sperrylite ($PtAs_2$) - postmagmatic, hydrothermal metasomatic association of PGE mineralization (on the right).

4 Conclusions

Thus, based on the identified features of the noble metal mineralization of the ore horizon "C", it can be divided into two types: low-sulfide platinum and properly platinum. The first type is characterized by a close relationship with sulfide minerals: noble metal minerals form intergrowths with the main sulphides, are included in rock-forming silicate minerals. This type is typical for the ore horizons of the Sungiyok area. Minerals of noble metals are localized in cracks and streaks and are associated with minerals formed during the post-magmatic stage of intrusive evolution. This type is observed in Sunghiyok area and Chuarvy area.

Acknowledgements. Authors are grateful to V.V. Shilovskich for the research of minerals which carried out on the "Hitachi S-3400 N" scanning electron microscope at "Geomodel" Resource Centre of Saint-Petersburg State University.

References

Subbotin VV, Korchagin AU, Savchenko EE (2012) Platinometal mineralization of the Fedorovo-Pana ore cluster: types of mineralization, mineral composition, genesis peculiarities. Vestnik Kola Sci Cent Russ Acad Sci 1:55 65

Voytekhovich VS, Kazanov OV, Kalinin AA (2008) Report on the results of prospecting and assessment work on platinum metal mineralization in the eastern part of the PanaTundras-massif in 2006–2008. LLC Kola Mining and Geological Company, Apatity

51

Genetic Problem of Quartz in Titanium Minerals in Paleoplacers of Middle Timan

I. Golubeva[1], I. Burtsev[1], A. Ponaryadov[1], and A. Shmakova[1,2](✉)

[1] Institute of Geology Komi SC UB RAS, Syktyvkar, Russia
alex.sch92@yandex.ru

[2] Karpinsky Russian Geological Research Institute (VSEGEI), Saint-Petersburg, Russia

Abstract. Titanium ore in the Devonian paleoplacers of Middle Timan is predominantly represented by leucoxene, less frequently by modified ilmenite (pseudorutile). Other titanium minerals are found in small amounts (or have a sharply subordinate significance). All titanium minerals have numerous inclusions of quartz, which create an intractable problem in the enrichment of titanium ore. Metamorphogenic porphyroblastic explains the presence of quartz in titanium minerals. Precambrian seric-chlorite clay weathering crusts are a supplier of titanium minerals Timan. Leucoxene and ilmenite form in paraschist under conditions of facies of green shale of regional metamorphism, poikiloblasts. In the poikiloblasts, the poikilite and helicitic structures are well defined, due to numerous poikilite inclusions of quartz. The poikiloblasts, poikilite and heliic structures are well represented, due to the numerous poikilite incorporating quartz.

Keywords: Titanium · Paleoplacer · Shale · Poikiloblastez · Leucoxene · Ilmenite

1 Introduction

The main resources of titanium in Russia are concentrated in identified (Pizhemskoe and Yaregskoye deposits) and are designed (Vodnenskoe and other manifestations) of the Devonian titanium paleoplacers Timan. Ores are represented by leucoxene, to a lesser extent altered under exogenous conditions by ilmenite (pseudo-ethyl). Ore is difficult to enrich due to the large number of quartz inclusions in titanium minerals. A large number of quartz inclusions in titanium minerals is an exceptional feature of Timan paleoplacers. The high content of quartz in titanium minerals is explained by its primary metamorphogenic genesis. Slates of the Precambrian folded basement of the Timan are the root source of titanium minerals. Titanium paleoplacers are formed due to the redeposition of the weathering crust on shale (Kochetkov 1967; Kalyuzhny 1972; Makhlaev 2006; Ponaryadov 2017).

2 Methods and Approaches

Photos of minerals were taken on a JSM-6400 scanning microscope with a Link ISIS-300 energy-dispersive spectrometer and a polarized microscope.

3 Results and Discussion

Titanium minerals - ilmenite and leucoxene (rutile, anatase and quartz aggregate) crystallize the parachale under the conditions of regional metamorphism of the green slate facies. This process is widely developed in the Riphean schists of Timan. The content of titanium minerals varies from 1.5–3.0%, occasionally rising to 5%. Titanium minerals crystallize in the form of porphyroblasts saturated with numerous quikilic poikilite inclusions. Poikilite inclusions of quartz are fragments of aleurite sizes. The quikite and inclusions of quartz and sericite captured during porphyroblasty determine the helicocyte structure (Fig. 1a, b). Titanium minerals are easily separated from shale and are separated during physical and chemical weathering. They accumulate due to gravitational separation during transportation and redeposition of weathering products. In paleoplacer metamorphic structures of titanite in titanium minerals are well preserved (Fig. 1d, f). In titanium minerals, fragments of quartz veins recorded with the growth of titanium minerals are diagnosed in paleoplacer. Sometimes fragments of sericite-chlorite schists with ilmenite grains are found (Fig. 1e). The quartz inclusions in leucoxene upon lithification of the ore-bearing sandstone can be regenerated with an increase in the volume of inclusions. The amount of SiO_2 in titanium minerals in bedrock - shale is 6.6–11.47%, and in paleoplacer - 12.2–28.19% (Ignatiev 1997). The high content of silicon dioxide in the titanium minerals of paleoplacer Timan is a specific feature of ore in this area and is explained by the metamorphogenic porphyroblastic genesis. For comparison, the SiO_2 content in titanium minerals of another well-known Tuganov paleoplacer in Western Siberia is given: silica is 1.82% for

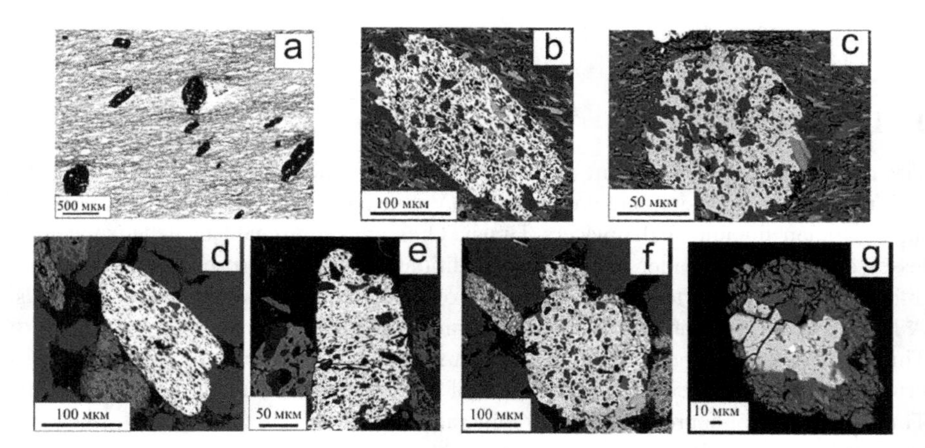

Fig. 1. Titanium minerals in shale and paleoplacers: a - leucoxene porphyroblasts in Riphean sericite-chlorite shale; b - helicitic structure in cross section of leucoxene porphyroblast in schist; c - poikilite inclusions in the longitudinal section of porphyroblast ilmenite in schist; d - helicitic structure in cross section leucoxene plate from paleoplacers; e - helicitic structure in leucoxene from paleoplacers; f - a longitudinal section of ilmenite with poikilitite inclusions of quartz from paleoplacers; g - a rounded fragment of sericite-chlorite shale with the inclusion of ilmenite from paleoplacers

ilmenite and 8.5.4% for leucoxene 6.5. The primary metamorphogenic nature of titanium minerals in Timan is indirectly indicated by the same increased content of MnO in porphyroblasts of leucoxene and ilmenite in native shale (2.5–3.63%) and in paleoplacer (to 2.33%).

4 Conclusions

The presence of a large amount of quartz in the titanium minerals of the paleoplacer Timan is explained by the porphyroblastic growth of shales under conditions of regional metamorphism of ilmenite and leucoxene. The increase in the percentage of silica in titanium minerals is also due to the processes of lithogenesis.

Acknowledgements. This research was supported by project AAAA-A17-117121270037-4 "Scientific basis for effective development and use of the mineral resource base, development and implementation of innovative technologies, geological and economic zoning of the Timan-North Ural region".

References

Ignatiev VD, Burtsev IN (1997) Timan's leucoxen: mineralogy and technology issues. Science, St. Petersburg 215 p

Kalyuzhny VA (1972) Geology of new placer formations. Science, Moscow 263 p

Kochetkov OS (1967) Accessory minerals in the ancient strata of Timan and Kanin. Science, Leningrad 200 p

Makhlaev LV, Golubeva II (2006) Ilmenite-containing metapelites as the most important source of the formation of giant and supergiant titanium placers. Titanium-zirconium deposits of Russia and prospects for their development. Moscow, IGEM RAS, pp 39–42

Ponaryadov AV (2017) Mineralogical and technological features of ilmenite-leucoxene ores of Pizhemskoe deposit, Middle Timan. Vestnik of the Institute of Geology, Komi SC UB RAS, no 1, pp 29–36. https://doi.org/10.19110/2221-1381-2017-1-29-36. (in Russian)

Oil and Gas Reservoirs, Including Gas Hydrates

52

Oil and Gas Reservoirs in the Lower Triassic Deposits of the Arctic Regions of the Timan-Pechora Province

1 Introduction

52

Oil and Gas Reservoirs in the Lower Triassic Deposits in the Arctic Regions of the Timan-Pechora Province

N. Timonina(✉)

Institute of Geology Komi SC UB RAS, Syktyvkar, Russia
nntimonina@geo.komisc.ru

Abstract. The Lower Triassic sediments of the Timan-Pechora oil and gas bearing province were studied by the complex of lithological, petrophysical and geochemical methods. It is established that productive deposits are represented by various-grained sandstones that formed in arid climates in the vast alluvial-lacustrine plain. The research allowed to study and characterize the structure features of rock and mineral aggregates, forming a pore space. The high heterogeneity of the composition and structure of the cement minerals of the reservoir caused by local facies-paleogeographic sedimentation conditions, were the cause of the significant variability of the pore space. Rocks reservoirs are complex, with high content of clay component, an effective development of which requires special methods of stimulation.

Keywords: Oil and gas bearing province · Lithological types · Reservoir · Clay minerals · Pore space · Porosity · Permeability

1 Introduction

In recent years, researches were intensified in the field of the formation of natural reservoirs. The basis for sedimentological reconstructions lies in the idea that the morphology and filtration characteristics of natural reservoirs are largely predetermined by ancient sedimentation situations, which are closely associated with the tectonic history of the territories. The Timan-Pechora province occupies the region of the northeastern Russian platform. Hydrocarbon accumulation occurred in Triassic rocks largely in the northern part of the basin, where pools were found in the rocks in the Varandey, Toravey, Labagan fields of the Sorokin swell, Kumzhinskoe, Korovinskoe – in Denisov Depression. Triassic deposits are distributed almost throughout the Timan-Pechora oil and gas bearing province, with the exception of the axial zones of large positive structures. The Lower Triassic includes strata of the Charkabozhsky suite, which thickness varies from the first meters in the southwest (in the Seduyahinsky swell) to 380 m in the central part of the Kolvinsky megaswell, the Khoreyver depression, the average thickness of the suite is 150–250 m (Morakhovskaya 2000). The sediments are represented by rhythmic alternation of red-brown clays, greenish-gray siltstones and gray sandstones with conglomerates and gravelites. Sandstones are characterized by a variety of granulometric composition from fine to coarse-grained, as well as a wide range of textures: massive, diagonal and horizontal bedding.

2 Methods and Approaches

Oil and gas reservoirs are geological bodies consisting of reservoir beds, lenses, and reservoirs of weakly and impermeable rocks of intra-reservoir tires, forming a single hydrodynamic system, bounded below and above by inter-reservoir tires. Accumulation of hydrocarbons in the reservoir and their safety are determined by the quality of each of the elements. The main reason for the differentiation of natural reservoirs by properties is their formation. The structural features of sedimentary layers determine the patterns of distribution of collectors and seals in them, their interrelationship, and ultimately predetermine the morphology and properties of natural reservoirs.

This paper proposes a conceptual model of the formation of a natural reservoir confined to the deposits of the Charkabozhskaya suite. The construction of working models was carried out using the results of complex processing of all available information, including well-logging, and study of core material.

The results of this work are based on combination of analytical tools, including polarizing microscope, scanning electron microscope, X-ray diffraction and electron microprobe, were used in order to identify and estimate authigenic mineral type sand to determine paragenetic sequences.

3 Results and Discussion

At the first stage of creating a geological model of a terrigeneous reservoir, separate strata were correlated. The selection and tracing of the layers was carried out by logging diagrams; local surfaces associated with homogeneous rocks, more or less sustained over the area, were taken as main reference points; within individual areas, additional benchmarks were used, which were characterized by stable geophysical characteristics. Two productive formations were confined to the Lower Triassic deposits. To maximally take into account the features of the structure of productive layers at the first stage of modeling, the study of microscopic in homogeneities of productive sediments was carried out on the basis of lithofacies and sedimentation models.

The main reason for the heterogeneity of natural reservoirs in terms of the properties of the reservoirs and tires that form them is the conditions of their formation. Detailed lithofacies analysis of Lower Triassic deposits was carried out, based on a set of investigations by both domestic (Muromtsev 1984; Bruzhes 2010; Morozov 2013) and foreign researchers (Celli 1989; Hellem 1983). The analysis of geological and geophysical information, the study of well cores, and the interpretation of well logging data allowed reconstructing conditions for the formation of lower Triassic natural reservoirs. The formation of these sediments took place in the continental conditions of the alluvial plain.

According to classification by A.G. Kossovskaya and M.I. Tuchkova (Kossovskaya, Tuchkova 1988) sandstones fall into the field of polymictic (SiO_2 content 62–78%) and volcanictic (SiO_2 content 54–64%). According to classification of Pettijohn (1976) points of the composition of sandstones are localized in the fields of graywacke.

Dominant cement minerals in Triassic sandstones include calcite, kaolinite, smectite, illite, and chlorite. Carbonate cement include pure calcite and siderite. The clear transparency and delicate crystalline habit of most of the clay, as revealed by microscope and SEM analysis, leads us to conclusion that the most clays in these sandstones are authigenic. Kaolinite occurs mostly as pore-lining and pore-filling cement, some of them appear as visible alteration of detrital feldspars. The distribution of kaolinite is uneven both in the section and in area. In the transition from coarse to fine-grained sediments, a decrease in the content of kaolinite is noted. Chlorite is common as pore-filling and lining cement. Smectite occurs both as grain-coating and filling pores. The studied deposits are characterized by a wide distribution of smectite minerals with increasing content upward the section. In the fine-grained sandstones of the basal stratum, its content does not exceed 50–60%, whereas in the upper part of the section its amount increases to 80–90%. The distribution and amount of clay minerals in sandstone cement is determined by both conditions of sedimentation and post-sedimentation transformations. The most widespread collectors are of III–V classes according to the A.A. Khanin classification (Khanin 1976). Class VI reservoirs are characteristic of floodplain formations and represented by aleurolites and fine-grained sandstones, in which large pore channels are practically absent. Class V collectors are represented by fine-grained sandstones with pore cement predominantly of smectite composition, low values of filtration properties are due to the insignificant content of large pore channels (less than 5%) and increasing number of non-filtered pores. These formations were deposited in floodplain conditions. The class IV reservoirs include fine- and medium-grained sandstones with polymineral cement: clayey pore and carbonate clot-pore type.

Collectors of II–III classes are represented by coarse-grained and medium-grained poorly sorted sandstones with pore-type cement, formed in an environment with a relatively quiet hydrodynamic regime.

4 Conclusions

The analysis of the reservoir properties of sandstones shows that sandstones formed under channel conditions are characterized by high median and average values of porosity and permeability, respectively 24% and 20.8% (porosity), $56 \cdot 10^{-12}$ and $40 \cdot 10^{-12}$ m^2, the lowest values characterize sediments formed in floodplain conditions: the median values of porosity and permeability do not exceed 11% and $0.6 \cdot 10^{-12}$ m^2, arithmetic averages reach 14% and $1.6 \cdot 10^{-12}$ m^2. Analysis of the graphs of the dependence of porosity-permeability for productive layers showed that the highest values of filtration properties were typical to deposits of the basal layer, where I–II class reservoirs were identified.

A reliable seal of subregional distribution is the deposits of the Upper charkabojskaya sub-suite, having high quality and thickness. Minerals of the smectite dominate in the mineral composition, chlorite and illite are also present. Middle-Upper Triassic clays are floodplain and lake originated, in their composition chlorite and illite predominate, which reduces the quality of the foamed seal compared to the Charkabozhskaya seal.

The differences in the composition and type of cement requires an individual approach, a balanced choice of technologies in determining the development strategy of fields and careful selection of a set of methods aimed at increasing oil recovery for different sections of fields.

Acknowledgements. The article was created under partial financial support by project of Ural Branch of Russian Academy of Sciences № 18-5-5-13 "The Geological Models, Environmental Conditions and Prospects of Oil and Gas bearing of Phanerozoic deposits in Arctic regions of Timan-Pechora province".

References

Bruzhes LN, Izotov VG, Sitdikova LM (2010) Litofacial conditions of formation of the horizon of the Yu1 Tevlinsko-Russkinskoye field of the West Siberian oil and gas province. Georesources 2(34):6–9

Celli RCh (1989) Ancient sedimentation. Nedra, Moscow, p 294

Hellem E (1983) Interpretation of facies and stratigraphic sequence. Mir, Moscow, p 328

Khanin AA (1976) Petrophysics of oil and gas reservoirs. Nedra, Moscow, p 259

Morakhovskaya ED (2000) Trias of the timan-ural region (reference sections, stratigraphy, correlation). Biochronology and correlation of the Phanerozoic oil and gas basins of Russia, vol 1, p 80. SPb: VNIGRI

Morozov VP, Shmyrina VA (2013) Influence of secondary changes in reservoir rocks on reservoir properties of the БС111 and ЮС11 productive layers of the Kustovoye deposit. Uchenye zapiski Kazan University, Kazan, vol 155, pp 95–98

Muromtsev VS (1984) Electrometric geology of sand bodies - lithological traps of oil and gas. Nedra, p 260

Pettijohn FJ, Potter PE, Siever R (1976) Sand and sandstone. Mir, Moscow, p 536

53

A Bench Scale Investigation of Pump-Ejector System at Simultaneous Water and Gas Injection

S. Karabaev(✉), N. Olmaskhanov, N. Mirsamiev, and J. Mugisho

Department of Mineral Developing and Oil and Gas Engineering, Engineering Academy, RUDN University, Moscow, Russia
simpleforfiza@mail.ru

Abstract. In this paper, a bench study of the pump-ejector system for simultaneous water and gas injection (SWAG) was conducted. For these purposes, a pump-ejector system stand was used. A differential pressure gauge was used to determine the gas flow at the ejector intake. According to the results of differential manometer calibrations, a new formula was obtained which reduces its inaccuracy to 1% at pressures below 0.6 MPa. In addition, according to the pressure-energy diagrams, it was determined that the gas injection with excess pressures in the ejector suction chamber significantly increases the efficiency of the pumping-ejector system overall.

Keywords: Liquid-gas ejector · Pump-ejector system · Liquid-gas mixture · SWAG

1 Introduction

Nowadays, a global trend of oil fields with increased residual oil saturation is observed (Drozdov and Drozdov 2012). There is also problem with associated petroleum gas (APG) utilization, which is not used rationally and is directed to gas flares. Currently, one of the most promising areas in oil production is the simultaneous water and gas injection (SWAG) technology. The definite advantage of the SWAG is the ability to use APG as a liquid-gas mixture compound. This method makes it possible to mount simple maintaining, reliable and efficient equipment, while providing a significant reduction in power consumption and increasing the SWAG efficiency (Drozdov et al. 2012).

2 Methods and Approaches

The investigation was conducted based on bench tests. The electric centrifugal pump and the liquid-gas ejector were used, which represent a single bench-model of the pump-ejector system. This stand is designed to investigate the characteristics of model ejectors, multistage centrifugal pumps and pump-ejector systems with liquid-gas mixtures using fresh water as a liquid, and air as a gas. The differential pressure gauge was used to determine the gas flow, with allowance for the excess pressure created in the suction chamber of the ejector. For operations at low pressures, the differential pressure gauge was pre-calibrated before.

3 Results and Discussion

Calibration of the differential manometer made it possible to clarify the existing formula for determining the gas flow rate at the ejector suction chamber at pressures below 0.6 MPa. The inaccuracy of the calculated values obtained by this method was less than 1%. According to the results of the conducted research, the pressure-energy characteristics of the ejectors were constructed during gas suction by the liquid. There occurs a significant expansion of liquid-gas ejector working range by the injection of gas with excess pressure in the ejectors suction chamber as well as the areas of maximum values of injection coefficients and efficiency rates are shifted to higher region of operating pressure.

4 Conclusions

Further study of the pressure-energy characteristics behavior and the causes affecting this change will facilitate the increase of efficiency of technological processes which use the pumping-ejector systems.

References

Drozdov AN, Drozdov NA (2012) Laboratory researches of the heavy oil displacement from Russkoye field's core models at the SWAG injection and development of technological schemes of pump-ejecting systems for the water-gas mixtures delivering. In: SPE 157819, presented at the SPE heavy oil conference canada held in Calgary, Alberta, Canada, pp 12–14

Drozdov NA, Drozdov AN, Malyavko EA (2012) Investigation of SWAG injection and prospects of its implementation with the usage of pump-ejecting systems at existing oil-field infrastructure. In: SPE 160687, presented at the SPE Russian oil and gas exploration and production technical conference and exhibition held in Moscow, Russia, pp 16–18

54

Innovative Technology of using Anti-Sand Filters at Wells of the Vankor Oil and Gas Field

K. Vorobyev(✉) and A. Gomes

Department of Mineral Developing and Oil & Gas Engineering,
Engineering Academy, RUDN University, Moscow, Russia
k.vorobyev98@mail.ru

Abstract. In this article, the authors considered an innovative technology for restricting sands to reduce complications and watering during the development of the Vankor gas and oil field.

Keywords: Sand control filters · Waterflood · Well · Borehole zone · Formation pressure drop

1 Introduction

The development of oil and gas fields in complex mining and geological conditions is associated with various complications - a drop in reservoir pressure, an increase in the water content of the crude product and the amount of mechanical impurities, destruction of the integrity of the well bottom zone. Analysis of the sources (Vorob'ev et al. 2017) shows that sand manifestations are a multifactorial and multi-element sophisticated technical system.

The task of managing the sand formation processes includes such elements as sand forecasting and effective methods of influencing the sand manifestation phases in order to reduce negative effects.

2 Methods and Approaches

Studies confirm that during the removal of rock particles from the reservoir in the operation of wells in the bottom zone form the high permeability channels of various widths and lengths of fractures along and bedding planes, by which the bulk of gas and produced water is filtered (Ozhogina et al. 2017).

When considering the performance of geological and technical measures at the stockwell of OJSC Gazprom (Lyashenko et al. 2018) in the direction of technology for the elimination of sand production and the removal of wells from inactivity, the causes and factors of sand production are given: poorly cemented formation; the viscosity of the formation fluid; the velocity of fluid particles in the reservoir.

3 Results and Discussion

Pilot tests of filters of various designs at the Vankor oil and gas field. During 2018, pilot industrial tests of filters of various designs were carried out: slotted and multi-layer mesh.

The filter descends on an unremovable packer into the perforated interval zone and is placed opposite the entire interval in order to minimize the skin effect. The well is operated using a high-rate installation of an electric submersible pump.

At the base of a 168 mm column, filters with a total length of about 200 m are lowered into the perforation intervals. Standard filter life is 24 months. Behind the filter, another is formed - a natural filter.

The organization of the reservoir pressure maintenance system at the Vankor oil and gas field involves the use of both produced water and water from water wells.

4 Conclusions

According to the results of the analysis of the geological and technical conditions of water wells and the characteristics of the aquifers of the Vankor oil and gas field, it was decided to use downhole anti-sand filters to reduce the effect of removal of mechanical impurities. The results of the pilot-scale industrial tests were considered successful, which gave rise to the decision to implement filters on an industrial scale. As of January 2019, all water wells in the Vankor oil and gas field are equipped with anti-sand mesh and slot-type filters

References

Lyashenko V, Vorob'ev A, Nebohin V, Vorob'ev K (2018) Improving the efficiency of blasting operations in mines with the help of emulsion explosives. Min Mineral Deposits. 1:95–102

Ozhogina EG, Shadrunova IV, Chekushina TV (2017) Mineralogical rationale for solving environmental problems of mining regions. Gornyi zhurnal. 11:105–110

Vorob'ev A, Chekushina T, Vorob'ev K (2017) Russian national technological initiative in the sphere of mineral resource usage. Rudarsko Geolosko Naftni Zbornik. 2:1–8

55

Integrated use of Oil and Salt Layers at Oil Field Development

V. Malyukov and K. Vorobyev(✉)

Department of Mineral Developing and Oil & Gas Engineering, Engineering Academy, RUDN University, Moscow, Russia
k.vorobyev98@mail.ru

Abstract. Combined development of oil and salt layers of the oil field allows to obtain a mineralized solution to intensify the extraction of oil from the reservoir and create an underground reservoir in the salt reservoir for underground storage of hydrocarbons, including the creation of an underground gas storage (UGS) in rock salt.

Keywords: Oil reservoir · Salt formation · Field · Combined use

1 Introduction

The complex development of deposits and the most complete use of the resource potential of the developed deposits, as well as the multifunctional use of the waste space of mineral deposits, is one of the main tasks of mining science (Vorob'ev et al. 2017).

In a number of oil fields, the screen is a salt layer of various capacities located above the productive oil reservoir, which can be used to produce mineralized water with subsequent injection of mineralized water into the productive oil reservoir, and the production-capacity (underground reservoir) formed in the salt reservoir can serve as a storage of hydrocarbons extracted from the productive reservoir (natural gas, associated oil gas, oil).

2 Methods and Approaches

To obtain mineralized water, it is advisable to drill wells on the salt formation and conduct underground dissolution of rock salt to obtain a solution of rock salt of a certain concentration for injection through injection wells into the productive formation and extraction of oil from the productive formation through producing wells.

Pumping more fresh water into the clay-containing collector than the reservoir water reduces the permeability of the collector and makes it a low-permeable collector. Controlling the mineralization of injected water and the properties of clays in productive formations can significantly increase the oil recovery rate.

3 Results and Discussion

The use of mineralized water can reduce the hydration of formation clays, but it is desirable to select the composition of water that is most compatible with the formation components of the productive formation.

The use of mineralized water obtained by dissolving the salt layer of the oil field can largely solve the problem of compatibility of pumped mineralized water with reservoir water and mineral composition of the reservoir (Lyashenko et al. 2018).

When developing a salt formation with a capacity of several tens of meters with the supply of solvent through the drilling well, according to the technological regulations, an underground production is created-a container in rock salt (vertical or horizontal) for storage of petroleum gas along the way.

4 Conclusions

Complex use of oil and salt layers of the oil field allows to obtain a mineralized solution to intensify the extraction of oil from the reservoir and to create an underground reservoir in the salt reservoir for underground storage of associated petroleum gas (APG), i.e. to create an underground gas storage (UGS) in rock salt.

At Talakan oil and gas condensate field (Republic of Sakha) the implementation of combined development of oil and salt layers of oil and gas condensate field with the creation of an underground gas storage in rock salt and the use of mineralized solution for the intensification of oil extraction was started.

References

Lyashenko V, Vorob'ev A, Nebohin V, Vorob'ev K (2018) Improving the efficiency of blasting operations in mines with the help of emulsion explosives. Min Miner Deposits 1:95–102

Vorob'ev A, Chekushina T, Vorob'ev K (2017) Russian national technological initiative in the sphere of mineral resource usage. Rudarsko Geolosko Naftni Zbornik. 2:1–8

56

Associated Petroleum Gas Flaring

A. Vorobev[1,2(✉)] and E. Shchesnyak[2]

[1] Atyrau University of Oil and Gas, Atyrau, Kazakhstan
fogel_al@mail.ru

[2] Peoples' Friendship University of Russia (RUDN University), Moscow, Russia

Abstract. The article analyzes the current state and prospects for utilization of a hydrocarbon component dissolved in oil and released during its extraction and preparation - associated petroleum gas (APG). The authors studied the properties, characteristics and component composition of APG.

The analysis of the APG use at the international and regional levels is carried out. The main causes of flaring were discussed and the shortage of production capacities for APG processing in the Russian Federation was noted as one of the main factors in the high level of APG flaring in the country.

The paper notes possible ways of utilization of associated petroleum gas, which depend on oil production conditions, such as field characteristics, oil/gas ratio (gas-oil factor), and market opportunities for recovered gas. An overview of all APG utilization methods are presented, which focuses on unit costs, economic benefits and environmental impact reduction. The authors analyzed the innovative experience of effective APG use in the USA and Canada. Special attention is paid to the need to solve the problem of the effective use of APG in the Russian Federation, especially the reduction of its burning in flare plants.

Keywords: Associated petroleum gas · Associated petroleum gas utilization · APG flaring · Environmental pollution

1 Introduction

Associated petroleum gas (APG) is a kind of natural gas that is in oil deposits, either dissolved in oil or as a free "gas cap" over oil into the deposits. Regardless of the source, as soon as it separates from crude oil, it usually exists in mixtures of other hydrocarbons such as ethane, propane, butane and pentane; in addition, APG contains water vapor, hydrogen sulfide (H_2S) and carbon dioxide (CO_2), nitrogen (N_2) and other mixtures. Associated petroleum gas containing such impurities so it can not be transported and used without purification, as it is extracted in the process of oil production (Kartamysheva et al. 2017).

The volume and composition of APG depends on the area of production and on the specific properties of the field. In the process of extraction and separation of one ton of oil, it is possible to obtain from 25 to 800 m^3 of associated gas. Some of this gas is used or stored, because governments and oil companies have made significant investments for its extraction. However, individual companies burn APG because of technical, regulatory or economic constraints. As a result, thousands of flare stacks from more than 17,000 oil production facilities around the world burn about 140 billion cubic meters of natural gas per year, resulting in more than 350 million tons of CO_2 and a large variety of pollutants, including very dangerous.

The overall increase in global flaring compared to previous years is largely due to negative developments in only a few countries: Iran, Russia and Iraq. Satellite data show an increase in flaring in Iran by more than 4 billion m^3, in Russia by almost 3 billion m^3 and more than 1 billion m^3 in Iraq. Flaring in Russia is close to the global average compared to oil production; in the other two countries, flaring intensity is higher. This is the expenditure of a valuable energy resource that can be used to promote the sustainable development of producing countries. Thus, 149 billion m^3 of associated petroleum gas, burned in 2018, could turn into 750 billion kWh of electricity, which exceeds its total annual consumption by all countries of the African continent.

According to official data, the volume of extracted APG in Russia increased by more than 7% - to 65 billion m^3 in 2010 and over 70 billion m^3 in 2018. The immediate impact on the increase in the volume of recoverable APG was made by the growth of oil production in new areas, including the East Siberian fields.

2 Methods and Approaches

For a long time, oil companies simply burned this unwanted by-product. Its flaring requires a significant portion of the security system.

The term "gas flaring" indicates a gas combustion (without energy recovery) in an open flame, which is continuously lit on top of the flare stacks in the field of oil production (Knizhnikov et al. 2017).

Flaring occurs for three main reasons:

- emergencies: limited incineration for reasons of safety for short periods of time can always be necessary even after connecting the gas gathering pipeline;
- deficiency of gas utilization capacity - isolated well flaring: if the well starts to produce oil and gas without connecting to gas gathering pipeline or other gas utilization technology, the gas can be shut off;
- lack of gas utilization capacity - well burning through the pipeline: if the well is connected to gas gathering pipeline, but these systems can not process all gas from the well (due to lack of power or compression), some or all of the associated gas from the well can be flared.

Billions of cubic meters of natural gas are burned at oil production sites around the world. Gas combustion is a costly energy resource that can be used to support economic growth and progress (Vorob'ev et al. 2017).

Since 2012, the US National Oceanic and Atmospheric Administration and the Global Gas Flaring Reduction Partnership have begun to apply a method for estimating the volume of APG flared. This method consists in the use of satellite observational data in the visible and near infrared ranges.

The conclusion drawn from the research results is that the volume of flared APG in the world increased to 147 billion m^3 in 2015, compared with 145 billion m^3 in 2014 and 141 billion m^3 in 2013. According to data for 2015, Russia led this "anti-rating", burning 24 billion m^3 of APG, followed by Iraq (17.5 billion m^3), Iran (16 billion m^3) and the United States (8 billion m^3). Russia is also a "leader" (in third place after the USA and Canada) with 1,814 flare stacks, which burn APG.

At present, there are other possible ways of utilization of associated gas, alternative to flaring. Among them it is necessary to single out the following:

- re-injection of APG into oil reservoirs to maintain pressure and increase oil recovery (as a method of increasing oil recovery), or for possible conservation of it as a resource and use in the future;
- the use of gas as an energy source for production site or at oil producing facilities in the vicinity;
- the most effective way of utilization of associated petroleum gas is its processing at gas processing plants to produce dry stripped gas (SOH), a wide fraction of light hydrocarbons (LH), liquefied natural gas (LNG) and stable gasoline (SG).

Below is an overview of all APG utilization methods, which focus on unit costs, economic benefits and environmental impacts.

3 Results and Discussion

The indicator of its useful use has remained stable since the 2000s, within 73–79% of the total amount of extracted APG in the country. Only in 2014–2017, according to the public accounts of companies, it rose to 85–86%. According to the statement of representatives of government organizations, the indicators of productive processing of APG amounted to 90% in 2018.

The amendments to the law "On environmental protection" (№ 219) adopted in July 2014 caused such as significant reduction in the share of associated gas combustion, according to these amendments, the company is obliged to establish its technological standards at the level of application of the best available technologies. The total investment in increasing the useful use of APG was estimated at 200 billion rubles. According to the Ministry of energy of the Russian Federation, it is expected that the target of 95% of the associated gas will be used by the end of 2020 (Vorobyev et al. 2018).

Foreign experience of utilization of APG shows that flaring of gas in torches has decreased slightly over the past two years, and oil production has also declined. In particular, Nigeria reduced volume APG combustion to 8 billion m^3 of nearly 18% compared with 2013 year. The volume of associated gas flaring in the USA decreased from 11 billion m^3 in 2016 to less than 9 billion m^3 in 2018 due to the use of a number of innovative low-volume technologies.

One of the innovative technologies for the production of liquefied natural gas with small volumes of associated petroleum gas is LNG Production (Production Natural Gas Liquids, LH-Pro). The "LH Pro" process combines dehydration, compression, cooling and conditioning, eliminating the need for expensive glycol and refrigeration systems.

Hydrate formation is excluded due to the thermal integration system. The technology was developed by ASPEN and is used in the US and Canada.

4 Conclusions

Thus, the most rational ways of utilization of associated petroleum gas in Russia, depending on the volumes of its extraction are:

- at small volumes - covering own energy needs;
- with increased volumes - electricity generation and primary processing of APG to produce lean dry gas (LDG) as fuel for the boiler room and light hydrocarbons (LH) for disposal to the oil collector;
- at resources from 50 to 150 million m^3/year - processing with obtaining LDG, as well as LH and electricity;
- with the amount of APG in excess of 150 million m^3/year, processing of LDG, and NGL is recommended.

References

Kartamysheva YeS, Ivanchenko DS (2017) Associated petroleum gas and the problem of its utilization. Molodoy uchenyy. № 25, pp 120–124

Knizhnikov AYU, Il'in AM (2017) Problems and prospects of associated petroleum gas use in Russia. 2017 WWF Rossii, Moscow, p 34

Vorob'ev A, Chekushina T, Vorob'ev K (2017) Russian national technological initiative in the sphere of mineral resource usage. Rudarsko Geolosko Naftni Zbornik 2:1–8

Vorobyev AE, Zhang L (2018) Analysis of production and consumption of associated petroleum gas in China. Bulletin of Atyrau Institute of oil and gas № 2(46), pp 137–142

Analytical Methods, Instrumentation and Automation

57

The Potential of Lacquer Peel Profiles and Hyperspectral Analysis for Exploration of Tailings Deposits

W. Nikonow(✉) and D. Rammlmair

Federal Institute for Geosciences and Natural Resources (BGR),
Hanover, Germany
wilhelm.nikonow@bgr.de

Abstract. Lacquer peel profiles are a valuable technique to preserve and study sedimentary structures and depositional features outside the field. The use of the lacquer fluid Mowiol® has shown to work well as a simple and rapid preparation technique for tailings material from a tailings heap in Copiapó, Chile. The combination of lacquer peels with 2D mapping techniques such as Hyperspectral Imaging (HSI) can provide important sedimentary information such as particle size distribution on a large scale with little effort, which becomes quantifiable due to continuous 2D information and modern image analysis. The presented relationship of element or mineral distribution with particle size can serve as a tool for targeted and more focused sampling for mineral exploration in tailings and tailings analysis in general, considering future selective mining for economic or environmental reasons.

Keywords: Lacquer profile · Sedimentology · Hyperspectral Imaging · Tailings exploration

1 Introduction

Creation of lacquer peels is an old, but rarely recognized technique enabling preservation of sedimentary structures of granular materials. The profiles are also known as sedimentary peels, sediment plates or lacquer profiles and are similar to the early description of soil monoliths by e.g. Voigt (1936). As a relatively simple and inexpensive technique with great possibilities in various geoscientific fields it represents a valuable method to study and preserve complex depositional processes and sedimentary textures on a scale from millimeters up to several square meters.

The general procedure is to apply the lacquer fluid on the target area, then let the fluid dry, solidify, remove the peel and fixate it on a board for transport and display. A detailed technical description of the preparation process can be found e.g. in Van Baren and Bomer (1979). The impregnation takes place due to gravity and capillary forces, which makes it possible to create vertical profiles.

The advantage of lacquer peel preparation is the possibility to preserve and transport features of interest for additional analysis on the site or in the laboratory. The combination of lacquer peels with emerging analytical techniques for 2D mapping such as Hyperspectral Imaging (HSI) can provide non-destructively valuable information from e.g. textural, sedimentological and geochemical analysis.

HSI is commonly used in remote sensing and finds its path into other fields of geoscience. HSI measures the reflectance of the samples in a wavelength range of, in this work, the visible and near infrared region (VIS-NIR: 400–2500 nm). It utilizes molecule - light interaction such as vibrational processes or electron energy level transitions which produce characteristic absorption features at certain wavelengths (Hunt and Salisbury 1970).

2 Methods and Approaches

The lacquer profiles were taken from a tailings heap in the city of Copiapó in northern Chile. According to historic aerial images it was deposited ca. 1993 on an area of ca. 230 × 130 m with an approximated mass of 400.000 t in 2018. There is only little information about the origin of the heap, but the material is most probably from a nearby processing plant using flotation or leaching to process ore from one of the many mines in the Punta del Cobre district around Copiapó. Since the heap is not covered, it is a target for erosion by wind and rain affecting the adjacent area including the Copiapó River.

Three lacquer profiles were taken from the steep parts of the eastern and western sides, where the internal layering becomes visible and accessible. To create the profiles the polyvinyl alcohol Mowiol® 4-88 was used. It is dissolved in hot water a day before preparation. At the site, the selected surface is scraped even and the Mowiol is applied on the surface by a spray bottle or a brush. For stabilization, several layers of Mowiol and gauze are laid on the back of the profile. Then the profiles are left to dry for about 24 h. Finally, they can be removed with a knife and, wrapped in cloth, can be transported easily. Preparation of one profile of about 30 × 50 cm takes about 30 min excluding the time for drying. One profile was taken from the eastern side (LP1) and two profiles from the western side (LP2, LP3). Additional samples from each layer were taken next to the profiles for geochemical, particle size analysis and verification of the HSI analyses.

The dry profiles were analyzed in the laboratory with the SisuRock system from Specim. The images were taken with two push broom cameras covering the wavelength range from 400 to 2500 nm (VNIR and SWIR) at a spatial resolution of 875 μm (SWIR) and 280 μm (VNIR). The data were processed using vertical de-striping and minimum noise fraction transform.

3 Results and Discussion

From the tailings impoundment three lacquer profiles were prepared. The profiles were left to dry for about 24 h to minimize the risk of destroying structures during removal and transport. When the profiles were removed from the tailings wall, only small areas at the edges or in clay layers were lost, but overall, the texture was well preserved.

Hyperspectral images of all profiles were obtained in the laboratory with the SisuRock system within minutes. The data show clearly the layered structure, which seems to be mostly an effect of particle size. Figure 1 shows the optical and the hyperspectral image of LP2. Layers that were differentiated optically in the field are indicated by capital letters. In the hyperspectral image areas of coarse particle size appear dark, while layers of fine particle size appear light grey or white. The data show a regression coefficient of $R^2 = 0.76$ at 860 nm between particle size and reflectance.

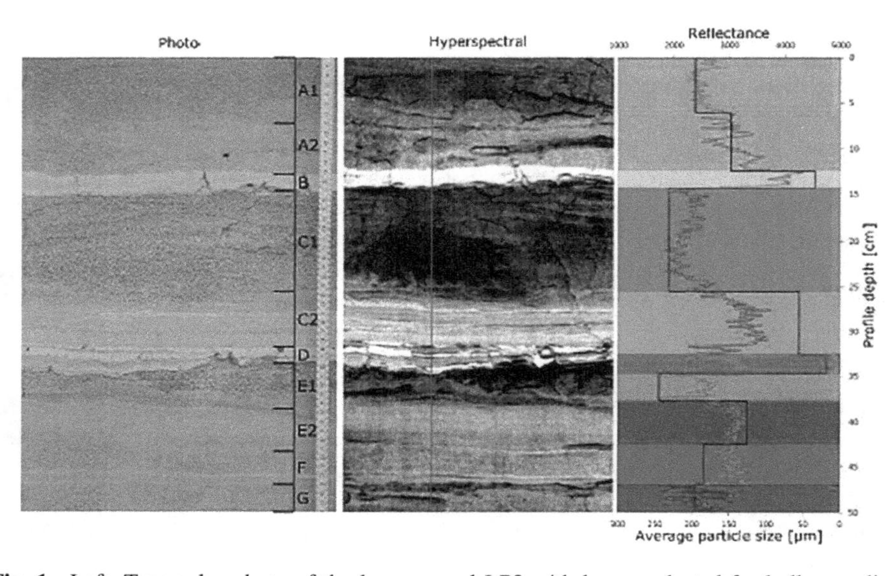

Fig. 1. Left: True color photo of the lacquer peel LP2 with layers selected for bulk sampling. Middle: Hyperspectral scan of LP2 at 860 nm. The red line indicates the position of the reflectance profile in the right graph. Right: Distribution of reflectance (red line) at 860 nm and average particle size (black line) at lacquer peel LP2.

The effect of particle size on reflectance in the visible and infrared wavelength region has been studied in the literature and was described as a negative correlation (Hunt and Vincent 1968; Okin and Painter 2004; Salisbury and Hunt 1968). However, the application on and implications for tailings evaluation still have to be described. The particle size distribution is an important information for the evaluation of the economic value of tailings deposits. The extractability of e.g. Cu depends largely on the particle size of the Cu minerals, since flotation or heap leaching depend on the access to free mineral surfaces. The geochemical data from the additional samples analyzed by XRF show a possible trend of increasing Cu concentration with increasing particle size above 100 µm. These areas of interesting particle size can be identified, located and quantified with HSI. After establishing the methods in the laboratory, it could be expanded to drill cores or in-situ measurements in the field. Therefore, the combination of creating lacquer profiles and HSI and image analysis serves as a rapid and simple tool to acquire important information for targeted and more selective sampling and even evaluation of tailings material for possible selective mining.

4 Conclusions

The proposed method of creating lacquer profiles has worked well on tailings material. It is a simple and rapid method to preserve sedimentary features for display and further analysis. The combination with 2D mapping techniques such as HSI provides rapidly continuous 2D information on sedimentary structures and particle size distribution on a scale from micrometer to meter. The information is quantifiable by image analysis and can be used for various geoscientific applications including sedimentology, sequential analysis and tailings exploration.

Acknowledgements. The results of this work are part of research that is funded by the German Federal Ministry of Education and Research (BMBF) within the projects SecMinStratEl (033R118B) and SecMinTec (033R186B). We thank Dominic Göricke for preparation of the lacquer profiles and Dr. Martin C. Schodlok for support with the hyperspectral data processing. We are very grateful to the reviewers for their helpful comments.

References

Hunt GR, Salisbury JW (1970) Visible and near-infrared spectra of minerals and rocks: I silicate minerals. Mod Geol 1:283–300

Hunt GR, Vincent RK (1968) The behavior of spectral features in the infrared emission from particulate surfaces of various grain sizes. J Geophys Res 73:6039–6046

Okin GS, Painter TH (2004) Effect of grain size on remotely sensed spectral reflectance of sandy desert surfaces. Remote Sens Environ 89:272–280

Salisbury JW, Hunt GR (1968) Martian surface materials: effect of particle size on spectral behavior. Science 161:365–366

Van Baren J, Bomer W (1979) Procedures for the collection and preservation of soil profiles. International soil museum

Voigt E (1936) Die Lackfilmmethode, ihre Bedeutung und Anwendung in der Paläontologie, Sedimentpetrographie und Bodenkunde. Zeitschrift der deutschen geologischen Gesellschaft 88:272–292

58

Thermometry of Apatite Saturation

Y. Denisova(✉), A. Vikhot, O. Grakova, and N. Uljasheva

Institute of Geology of the Komi SC UB of RAS,
Syktyvkar, Komi Republic, Russia
yulden777@yandex.ru

Abstract. The results of the study of accessory apatite from the Kozhym massif rocks have been presented in this paper. Apatites of the same morphological type were found in granites. The Kozhym massif granites formation temperatures by apatite were determined by the Watson and Bea saturation thermometry. These temperatures were compared with the previously obtained ones for accessory zircon of the same massif.

Keywords: Apatite · Granite · The Kozhym massif · The Subpolar Urals · Watson · Bea

1 Introduction

The Kozhym massif is located in the northeastern part of the Subpolar Urals on the left and right banks of the Kozhym River in the Oseu and Ponyu Rivers basins (Fig. 1) among deposits of the Puyvinian Middle Riphean Formation. This massif is the second by area among the geobodies composing the Kozhym intrusion which includes the Kuzpuayu granite massif. The most fully preserved Kozhym massif granites are medium-grained pink greenish-gray rocks. They (the rocks) are characterized by a massive coarse-platy texture with well-defined tectonic gneissiness. The studied massif belongs to the A-type according to B. Chappel classification (Fishman et al. 1968).

The accessory massif apatite is represented by yellow mat elongated crystals of a hexagonal dipyramidal-prismatic habitus. The crystals size is 0.1–0.4 mm, the elongation coefficient is 1.5–3. The mineral shape is represented by a combination of a prism (1010) and a dipyramid (1011). The faces surface is fractured. Characteristic inclusions are quartz, plagioclase, zircon.

2 Methods

Apatite is increasingly used as a geothermometer. The E. Watson saturation thermometry was used to determine the mineral crystallization temperatures. This method allowed determining the apatite and rock formation temperature by the distribution of the phosphorus oxide content between apatite and the rock containing the mineral. The level of phosphorus saturation necessary for the apatite crystallization depends on the

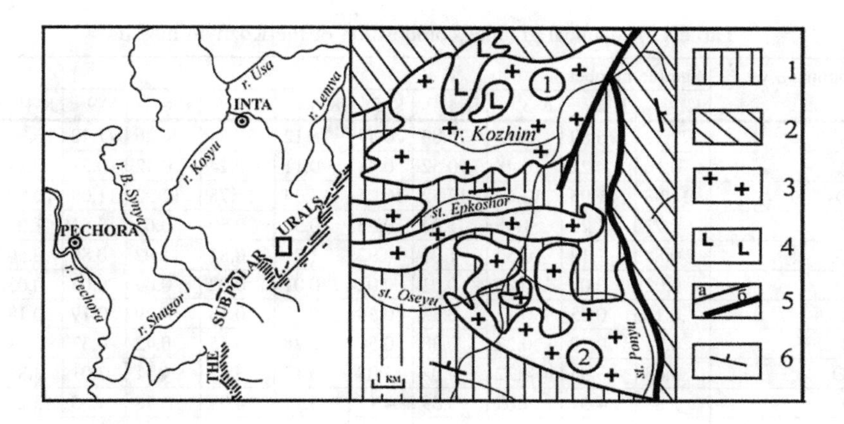

Fig. 1. The geological Kozhym granite massif map. 1 – mica-quartz schists, green orthoschists, quartzits; 2 – mica-quartz schists, porphyries, porphyrites, marbles and quartzits interlayers; 3 – granites; 4 – gabbro; 5 – contact lines: a – stratigraphic and magmatic, b – tectonic; 6 – planar structured bedding elements. Massifs (numbers in circles): 1 – the Kuzpuayu; 2 – the Kozhym.

silica content in the meta-aluminic rock (A/CNK < 1) and the temperature according to the Harrison and Watson calculations (Harrison et al. 1984):

$$\mathrm{InDp} = (8400 + 26400\,(\mathrm{SiO_2} - 0,5))/\mathrm{T} - 3,1 - 12,4\,(\mathrm{SiO_2} - 0,5),$$
$$\mathrm{P_2O_5(HW)} = 42/\mathrm{Dp},$$

where Dp is the ratio of P concentration in apatite and melt, P_2O_5; SiO_2 is the weight fraction of the phosphorus oxide, silica in the melt, wt. %, T is the temperature, Kelvin.

Bea (Bea et al. 1992) proposed the following addition to the Watson's formula for peraluminum rocks (A/CNK > 1):

$$\mathrm{P_2O_5(Bea)} = \mathrm{P_2O_5(HW)} * \mathrm{P_2O_5(HW)} * e^{\frac{6429(\mathrm{A/CNK}-1)}{T-273,15}}.$$

3 Results and Discussion

The apatite saturation temperatures based on the data of the chemical granites composition of the Kozhym massif (Table 1) were calculated by the author according to Watson and Bea (Table 2).

The Kozhym massif rocks formation temperatures distribution histograms by apatite were compiled on the obtained temperatures for each calculation method (Fig. 2).

Table 1. Chemical granites composition of the Kozhym massif

Component, wt. %	Sample number									
	K-1	K-2	K-3	K-4	K-5	K-6	K-7	K-8	K-9	K-10
SiO_2	77.78	76.89	75.95	75.89	76.49	78.12	77.54	76.26	77.48	76.95
TiO_2	0.16	0.22	0.48	0.52	0.48	0.11	0.24	0.42	0.31	0.59
Al_2O_3	11.88	11.95	12.69	12.52	10.05	11.34	11.78	12.22	11.09	10.92
FeO	1.72	1.29	1.15	1.24	0.56	0.50	0.59	0.62	0.61	1.03
Fe_2O_3	0.84	1.12	0.52	1.05	0.92	1.21	0.87	1.02	0.89	0.56
MnO	0.02	0.00	0.00	0.01	0.02	0.01	0.02	0.04	0.03	0.03
MgO	0.16	0.25	0.17	0.33	0.38	0.39	0.18	0.29	0.19	0.18
CaO	0.31	0.29	0.22	0.38	0.59	0.28	0.45	0.42	0.37	0.51
Na_2O	3.65	3.33	4.22	4.02	3.08	3.15	3.22	4.51	3.01	3.89
K_2O	3.88	4.51	4.09	3.89	4.15	5.17	4.99	3.78	4.65	3.28
P_2O_5	0.01	0.03	0.02	0.02	0.03	0.02	0.01	0.01	0.03	0.02
ппп	0.05	0.29	0.59	0.15	0.75	0.62	0.39	0.98	1.02	1.23
$\sum$	100.46	100.17	100.10	100.02	97.50	100.92	100.28	100.57	99.68	99.19

Note. The chemical composition was obtained using the silicate method in CUC ≪Science≫ of Institute of geology of Komi SC UB RAS (analyst Koksharova O.V.)

Table 2. Saturation temperatures for the Kozhymmassif apatite

Temperature, °C	Sample number									
	K-1	K-2	K-3	K-4	K-5	K-6	K-7	K-8	K-9	K-10
According to Watson	784	860	819	818	856	838	782	770	865	828
According to Bea	722	798	770	764	856	836	768	770	840	826

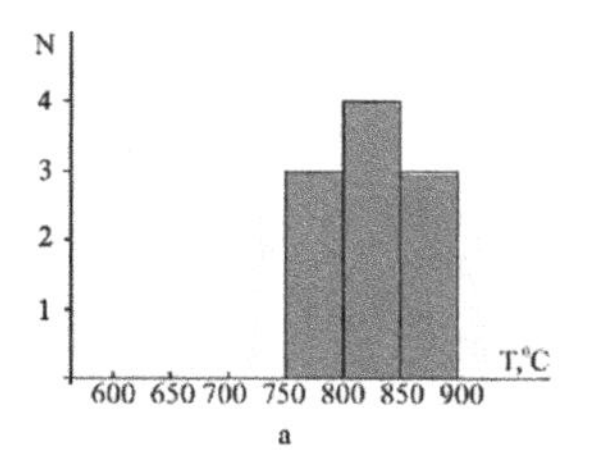

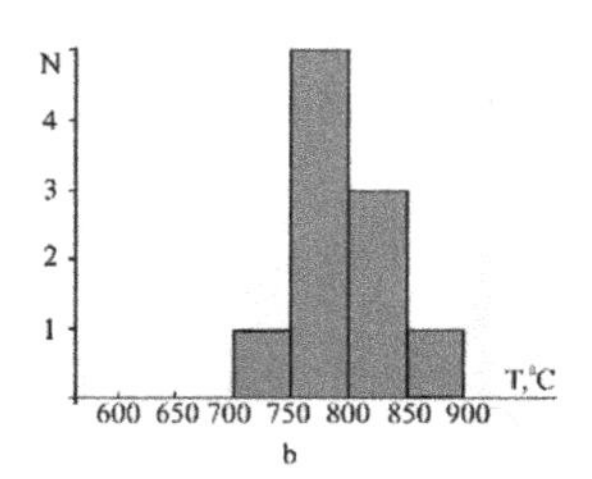

Fig. 2. Formation temperatures distribution histograms for the Kozhym massif granites. a - according to Watson, b - according to Bea

The presented histograms show that the studied granites are high-grade rocks. The Kozhym massif formation occurred at temperatures from 770 °C to 865 °C and averaged 822 °C according to Watson. The application of Bea refinements allowed concluding that the massif rocks formation occurred at temperatures from 722 °C to 856 °C and an average of 795 °C.

4 Conclusions

The Kozhym massif rocks formation occurred at high temperatures. The similar massif granites formation temperature ranges: 770–865 °C according to Watson, 722–856 °C according to Bea were obtained using the Watson saturation thermometry by apatite and the Watson formula adjustment for aluminous rocks according to Bea. The data confirmed the temperatures obtained earlier by the author using the Watson saturation thermometry (749–816 °C) and the classical evolutionary-morphological Pupin analysis (700–900 °C) for zircon of the same massif (Denisova 2016; 2018a; b). It can be affirmed that the Watson saturation thermometry and the refined Bea formula for apatite provide the same information about regime temperature evolution during the granites formation such as on the Watson saturation thermometry and the Pupin and Tyurko evolutionary-crystallomorphological analysis for zircons.

Acknowledgements. The work was supported by the Basic Research Program of RAS № 18-5-5-19.

References

Bea F, Fershtater GB, Corretgé LG (1992) The geochemistry of phosphorus in granite rocks and the effects of aluminium. Lithos 48:43–56

Denisova UV (2018a) Apatite of the Nikolaishor granite massif (the Subpolar Urals). Vestnik of Institute of geology of Komi SC UB RAS, Syktyvkar, № 9, pp 24–29

Denisova UV (2018b) Crystal morphology of zircon in solving problems of the Kozhimsky massif granites genesis (the Subpolar Urals). Trends in the development of science and education, №35, Part 3. Samara, pp 45–48. https://doi.org/10.18411/lj-28-02-2018-51

Denisova UV (2016) Temperature survey of zircon from the granitoids of the Subpolar Urals. Vestnikof Institute of geology of Komi SC UB RAS, Syktyvkar, №12, pp 37–44

Fishman MV, Yushkin NP, Goldin BA, Kalinin EP (1968) Mineralogy, typomorphism and genesis of accessory igneous rocks minerals of the Urals and Timan north. M.-L.: Sci 252

Harrison TM, Watson EB (1984) The behavior of apatite during crustal anatexis: equilibrium and kinetic considerations. Geochim Cosmochim Acta 48:1467–1477

Methods of Extraction of Micro- and Nano-Particles of Metal Compounds from Fine Fractions of Rocks, Ores and Processing Products

A. Smetannikov(✉) and D. Onosov

PFIC UB RAS "GI UB RAS", Perm, Russia

smetannikov@bk.ru

Abstract. The presented extraction methods are based on the features of the state of liquids in the capillary space in the form of a weak electrolyte. These methods make possible to extract micro- and nanoparticles adsorbed in matrix minerals from a suspension placed in the graphite substrate into a capillary solution. After the particle deposition in the substrate due to evaporative concentration the microprobe analysis is performed. The method is known as the capillary method of extracting micro and nanoparticles.

The described methods was used as a prototype for extracting micro- and nanoparticles from suspensions associated with the use of an external electric field. The field is created by connecting the electrodes to a graphite substrate and applied suspension with a direct current source using the voltage of $4 \div 6$ V. The micro- and nanoparticles adsorbed in matrix minerals are extracted into the capillary solution. The deposition of micro- and nanoparticles in a capillary solution is made by the method of evaporative concentration. The application of an external electric field intensify extraction of micro- and nanoparticles.

Keywords: Nanoparticles · Capillaries · Electrolyte · Graphite · Electric potential · Adsorption

1 Introduction

The following methods of extraction of minerals trace from rocks, ores and products of their processing are well known.

1. In heavy liquids with subsequent magnetic and electromagnetic separation of heavy fractions (Chueva 1950; Mitrofanov et al. 1974). The disadvantage of this method is the aggregation of nanoparticles with minerals with a particle size less than 0.045 mm and their absence in the heavy fraction.
2. Extraction of rhenium from ores of black shale formations (Oleynikova et al. 2012. The method for the extraction of micro- and nanoparticles of metal compounds is not calculated.
3. Extraction of nanoparticles from disperse systems (Zhabreev 1997) using the electric field created by electromagnets. The method does not provide for the extraction of all classes of metal compounds.

4. Method of capillary extraction of nanoparticles (Smetannikov 2014). The material of fine fractions (<0.25 mm) mixing with a liquid forms of suspension, placed in the form of a hemisphere in a graphite substrate. The distance between minerals is comparable to their size and forms a capillary space. The liquid acquires structure, charge and becomes electrolyte (Deryagin et al. 1989). Nanoparticles adsorbed in matrix minerals are extracted into the capillary solution.

Due to the evaporation concentration, the nanoparticles migrate to the base of the hemisphere. After drying ($1 \div 1.5$ h), the hemisphere is removed, and the nanoparticles on the substrate are examined under a microprobe (Fig. 1). More than 50 minerals have been identified: intermetallic compounds and solid solutions of Au, Cu, Ag, Zn, Pb, Ni, Sn, Cu, Fe, Cr, Ti zircon, Sn minerals, monazite, which are trace substances in the water-insoluble salt residues of the Verkhnekamsk deposit (Smetannikov and Filippov 2010).

The disadvantage of this method is incomplete extraction of nanoparticles. This method served as a prototype for developing method for extracting particles using the methods of intensifying the natural properties of capillary systems by an external electric field.

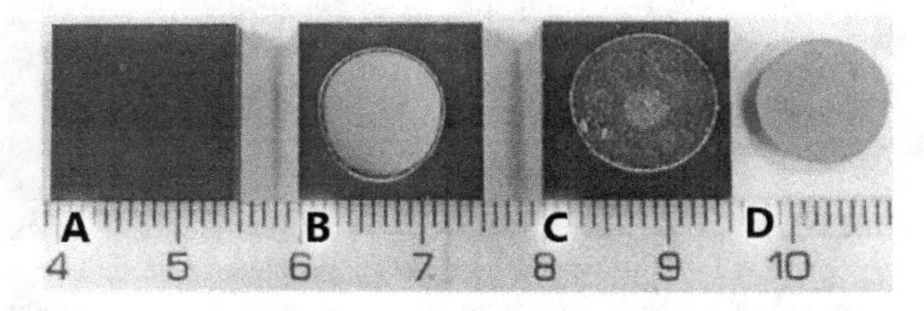

Fig. 1. A - clean graphite plate; B - a plate with a dried hemisphere of the investigated suspension; C- a plate with a hemisphere removed; D - Dried hemisphere (flat side up)

2 Methods and Approaches

The suspension is prepared from the material of fine fractions using distilled water and adding salt. The suspension is applied to a graphite plate, forming the hemisphere. An electrode with a "plus" sign (forming an anodic part of the system with a graphite plate) is connected to the underside of the plate. The second electrode is immersed in the suspension (without touching the graphite substrate) with a minus sign serves as a cathode. Then there is inclusion of the direct current voltage of $4 \div 6$ V. Further there is an electrolytic extraction of nanoparticles and deposition on the substrate. The dried hemisphere is removed and the sediment is analyzed.

There were conducted three experiences. The first is capillary sedimentation. The second experiment is anodic deposition, with a graphite plate serves as an anode. The third experiment is with pole reversal, when the upper (positive) electrode serves as the anode and the graphite plate as the cathode. Figure 2 posted three photos of deposition traces in three experiments.

In the first experiment, there were matrix minerals practically absent, the number of nanoparticles is minimal (Fig. 2A). In the second experiment, the number of matrix minerals and nanoparticles is 1–2 orders of magnitude higher than in the first experiment (Fig. 2B). In the third experiment, an intermediate result (Fig. 2C). The maximum of nanoparticles is fixed in the second experiment. Here, the deposition process is enhanced by heating the graphite plate.

Experiments have shown the natural properties of water in a capillary space change created by a constant current source under the influence of an electric field. Capillary solution acquires the properties of an electrolyte. The salt concentration in the suspension is 0.5–1%.

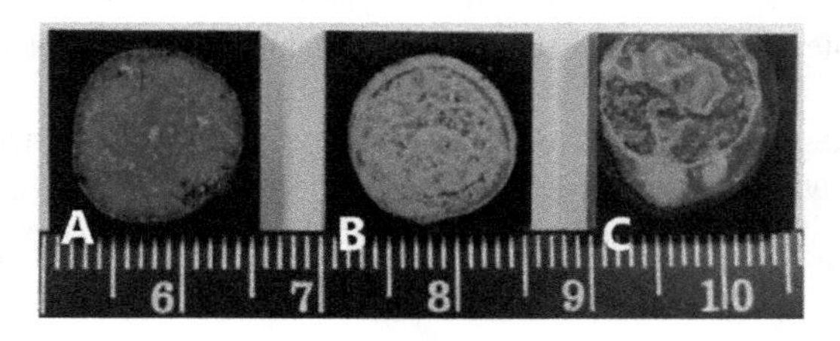

Fig. 2. A - sediment after capillary leaching; B - sediment after electrolytic anodic leaching; C - sediment after cathodic leaching

3 Results and Discussion

Extraction of micro- and nanoparticles from insoluble residues of salts, rocks, ores and products of their processing is provided by creating an electric field. The field creates the extracting properties of the electrolyte in suspension from the material of fine fractions and salt solution.

This factor ensures maximum extraction of micro- and nanoparticles and their fixation on a graphite substrate for subsequent analysis is by the microprobe method. Moreover, it is possible to solve the direct problem - the extraction of micro- and nanoparticles from insoluble residues of salts, ores and technogenic products, as well as the inverse problem - "cleansing" any small fraction of the material from impurities of minerals and metal particles.

While using the capillary-electrolytic method of extracting micro- and nanoparticles, the leaching effect (extraction) is achieved by 1–2 orders of magnitude higher than the results of the application of the prototype. The main goal is to obtain information of the form of finding elements-microimpurities. This result has been achieved. Researches are confirmed by the patent (Smetannikov and Onosov 2018).

4 Conclusions

1. The method relates to the methods for extracting micro- and nanoparticles of metal compounds from various media under the influence of direct electric current in a suspension of the material under study, water and salt, placed in a graphite substrate.

2. The new method is different from the prototype by the use of an external electric field, the intensity of extraction of nanoparticles and the completeness of extraction.

References

Chueva MN (1950) Mineralogical analysis of schlich and ore concentrates. Gosgeolizdat, Moscow, p 179

Deryagin BV, Churaev NV, Ovcharenko FD et al (1989) Water in disperse systems. Chemistry, 288 p

Mitrofanov SI et al (1974) Mineral studies for enrichment. Nedra, Moscow, p 352

Oleynikova GA et al (2012) A nanotechnological method for extracting rhenium from rocks and ores of black shale formations. Patent No. 245237 RU, St. Petersburg State University (RU), Claims 12 June 2010. Publ. 10 July 2012, Bul. No. 19

Smetannikov AF (2014) Capillary method of extracting micro- and nanoparticles of minerals from fine fractions for subsequent microprobe analysis. Yushkinsky readings 2014, Materials of the mineralogical seminar with international participation, Syktyvkar, pp 177, 178

Smetannikov AF, Filippov VN (2010) Some features of the mineral composition of salt rocks and products of their processing (for example, Verkhnekamsk salt deposit). Scientific readings of the memory of P.N. Chirvinsky: Sat. articles, Perm, Issue 13, pp 99–113

Smetannikov AF, Onosov DV (2018) Capillary electrolytic method of extracting micro- and nanoparticles of metal compounds from fine fractions of rocks, ores and industrial products. Patent №2659871, PFIC UB RAS (RU), Declared. 20 December 2016, Published 04 July 2018

Zhabreev VS (1997) Installation for the extraction of substances and particles from suspensions and solutions. Patent 32098193 (RU), Chelyabinsk State Technical University, Appl. 26 July 1995, Publ. 10 December 1997

60

Studies of Structural Changes in Surface and Deep Layers in Magnetite Crystals after High Pressure Pressing

P. Matyukhin(✉)

Belgorod State Technical University named after V.G. Shukhov, Belgorod, Russia
mpvbgtu@mail.ru

Abstract. The article introduces the study of structural changes in surface and deep layers in magnetite crystals, on samples under high pressure pressing. Magnetite (magnetite iron-ore concentrate) is widely used as a filling compound of new composites which are planned to be used in nuclear-construction. These compounds are based on the aluminum containing matrix with a filling compound. This modern composite material can be used in construction structures able to resist significant loads, operate in such extreme situations as abrupt dynamic loads, fires with further alternating temperature oscillations.

Keywords: Magnetite · Pressure · Crystal · Structure · Layer · Material

1 Introduction

Nowadays many scientists develop new kinds of materials, including composites which can be used in nearly all spheres of human life. Composite materials based on organic and inorganic components with different additives and fillers are created. These components can be the base for different matrixes. Fillers with different qualities are introduced into these matrixes depending on the purpose, scope of operation, conditions of service of the designed composite. These matrixes can have metal, ceramics, concrete, polymer and other bases, and the fillers may be cast iron shot, barites, metal processing remains, mining company products (for example, iron containing rocks) and of other enterprises (Grishina and Korolev 2016; Barabash et al. 2017; Matyukhin 2018; Gulbin et al. 2018; Kruglova 2009; Laptev et al. 2015; Gulbin et al. 2016; Matyukhin et al. 2011, Samoshin et al. 2017).

Composite materials based on metal matrixes can be used in bearing constructions production. Such constructions can resist high mechanical impacts, intensive ionizing radiation and alternating temperatures, that can widely be used in nuclear construction industry, and foremost, for biological protection on sites with power supplies of different nature. Nowadays, the issue of aluminum containing matrixes and iron containing fillers utilization in designing new types of radiation-protective construction composite materials becomes acute. One of promising fillers for new composite materials is magnetite. The introduction of modern composite materials of natural iron ore raw materials will give them additional mechanical strength, increase their radiation shielding characteristics. This filler is widely found in nature, is relatively cheap, and composites on its base will meet ecological requirements as they are a part of the ecosystem. Here a possibility of combining magnetite iron ore concentrate with aluminum containing matrix and receiving composite material by high pressure pressing is of great scientific interest. In this research it is necessary to solve issues about maximum possible filler compacting in its composite material matrix, studying composite filler behaviour at high pressing pressures, in particular studying structural changes in surface and deep layers of magnetite crystals at pressing pressures when such a composite material is produced, and compact-space position of its particles relative to each other.

2 Methods and Approaches

As an object of research we used a highly-dispersive enriched magnetite iron-ore concentrate from the Lebedinsk deposit of the Kursk Magnetic Anomaly with density 4950 kg/m^3, Mohs hardness – 6, black colour; the magnetite is in form of irregular needle-shaped grains, octahedral crystals, shell-like fracture; the mineral composition is introduced by magnetite with inclusions of carbonate (chalk-stone) and siliceous (quartz) admixtures. After cleaning and chemical treatment of the magnetite iron ore concentrate per 99,8% its mineral composition is introduced by magnetite ($FeO.Fe_2O_3$) with fractional composition of particles 0,05–15 mcm. In the research we used a modern electronic microscope “TESCAN MIRA 3 LMU”.

3 Results and Discussion

For experiment fairness the magnetite samples in the form of a high dispersive powder with particles 0,05–15 mcm were used, without adding it into the aluminium containing matrix. Figure 1 shows the structure of surface and deep layers of magnetite particles without pressure treatment (Fig. 1a) and after pressure pressing 5000 MPa (Fig. 1b) with magnifying power 10 mcm. The analysis of the received data showed that the surface and deep layers of magnetite crystal particles after pressure pressing at 5000 MPa (Fig. 1b), in comparison with the micro photo of magnetite crystal particles not pressed (Fig. 1a), there are spaces with loose distribution of its particles along the whole sample volume: there are numerous empty spaces; particles have irregular form with hard aggregation and rough edges. Figure 2 shows the structure of surface and deep magnetite particle layers, after pressure 10000 MPa and 20000 MPa with magnifying power 10 mcm.

a)

b)

Fig. 1. Magnetite surface with magnifying power 10 mcm: *(a)* without pressure pressing treatment, *(b)* after pressure treatment 5000 MPa

The received data analysis showed that at pressure pressing increase up to 10000 MPa (Fig. 2a) magnetite crystal particles were distributed more compactly within the whole volume of the studied sample: also there are numerous empty spaces but with a slight reduction of their geometrical sizes; the magnetite particles structure changed insignificantly, as before they had irregular form with strong aggregation and rough edges, but here particles of smaller fraction appeared, that evidenced the process of partial mechanical destruction of its particles.

a)

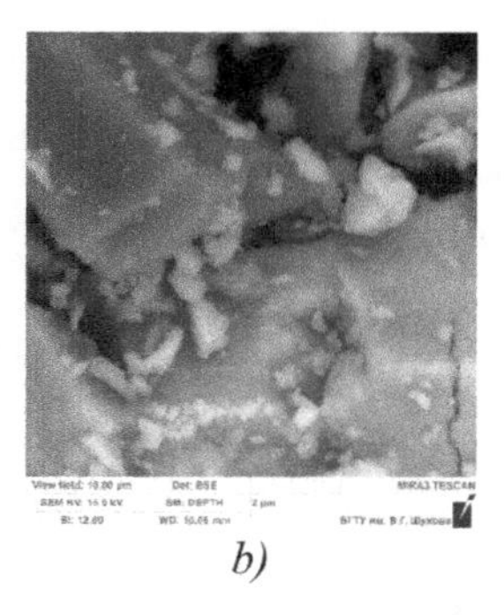

b)

Fig. 2. Magnetite surface with magnifying power 10 mcm: *(a)* after pressure pressing 10000 MPa, *(b)* after pressure pressing 20000 MPa

When the pressure increases up to 20000 MPa on the magnetite crystal particles (Fig. 2b), there is a more compact visual magnetite particle packing both in surface and deep layers of the studied sample; there is a significant increase of zones with good compaction. There is a greater degree of destruction of magnetite crystal particles surface in the total amount of material samples in comparison with previous pressure pressing; somewhere surface aggregation and hematite particles edges is smoothed.

It evidences qualitative distribution of hematite particles of all fraction composition along the whole sample volume, but micro photos show remaining zones with "insufficient compaction", where there is a lack of fine-fractioned magnetite particles.

4 Conclusions

Based on the studies of structure changes in surface and deep layers of magnetite crystals, after high pressure pressing and studied with electron microscope, we can make an assumption that the structure of the composite may be denser by adding magnetite into its filler 15–25% (by volume) of its particles, but smaller fraction in comparison the studied one. It is theoretically possible that after adding such an amount of magnetite fine-fraction a higher degree of sample material compacting will be achieved. Or to conduct the studies with adding the same percentage ratio into the volume of studied composite material matrix samples in the form of aluminum powder, as it has particles less in size than magnetite particles. We think that these studies have high theoretical and practical significance and they should be taken into account when developing new composite materials based on magnetite filler and different metal matrixes.

Acknowledgements. The work is realized in the framework of the Program of flagship university development on the base of the Belgorod State Technological University named after V.G. Shukhov, using equipment of High Technology Center at BSTU named after V.G. Shukhov.

References

Barabash DE, Barabash AD, Potapov YuB, Panfilov DV, Perekalskiy OE (2017) Radiation-resistant composite for biological shield of personnel. In: IOP conference series: earth and environmental science. C. 012085

Grishina AN, Korolev EV (2016) New radiation-protective binder for special-purpose composites. Key Eng Mater 683:318–324

Gulbin VN, Kolpakov NS, Gorkavenko VV, Boikov AA (2018) Research of the structure and properties of radio and radiation-protective polymer nanocomposites. J Electro-magnetic Waves Electron Syst 23(1):4–11

Gulbin VN, Martsenuk AV, Gorkavenko VV, Cherdyntsev VV (2016) Polymeric composites for radio and radio active protection. Sci Intensive Technol 17(10):7–12

Kruglova AN (2009) Radiation protective materials based on industrial wastes: physic-mechanical properties. Reg Archit Const 1:53–56

Laptev GA, Potapov Y, Yerofeev VT (2015) Development of manufacturing technology metalloconcretes. Build Reconst 1(57):123–129

Matyukhin PV (2018) The choice of iron-containing filling for composite radioprotective material. In: IOP conference series: materials science and engineering 11. International conference on mechanical engineering, automation and control systems 2017 - material science in mechanical engineering. C. 032036

Matyukhin PV, Pavlenko VI, Yastrebinskiy RN, Bondarenko YuM, (2011) Prospects of creating modern highly constructive radiation-protective metalocomposites. Bulletin of BSTU named after V.G. Shukhov. 2, 97

Ochkina NA (2018) Heat stability of radio-protective composite based on aluminous cement and polymineral industrial waste. Pridneprov Sci Bull 3(2):007–010

Samoshin AP, Korolev YV, Samoshina YN (2017) Internal stresses at metal concrete structure formation for protection from radiation. Bull. BSTU Named After V.G. Shukhov 6:13–17

PERMISSIONS

All chapters in this book were first published by Springer; hereby published with permission under the Creative Commons Attribution License or equivalent. Every chapter published in this book has been scrutinized by our experts. Their significance has been extensively debated. The topics covered herein carry significant findings which will fuel the growth of the discipline. They may even be implemented as practical applications or may be referred to as a beginning point for another development.

The contributors of this book come from diverse backgrounds, making this book a truly international effort. This book will bring forth new frontiers with its revolutionizing research information and detailed analysis of the nascent developments around the world.

We would like to thank all the contributing authors for lending their expertise to make the book truly unique. They have played a crucial role in the development of this book. Without their invaluable contributions this book wouldn't have been possible. They have made vital efforts to compile up to date information on the varied aspects of this subject to make this book a valuable addition to the collection of many professionals and students.

This book was conceptualized with the vision of imparting up-to-date information and advanced data in this field. To ensure the same, a matchless editorial board was set up. Every individual on the board went through rigorous rounds of assessment to prove their worth. After which they invested a large part of their time researching and compiling the most relevant data for our readers.

The editorial board has been involved in producing this book since its inception. They have spent rigorous hours researching and exploring the diverse topics which have resulted in the successful publishing of this book. They have passed on their knowledge of decades through this book. To expedite this challenging task, the publisher supported the team at every step. A small team of assistant editors was also appointed to further simplify the editing procedure and attain best results for the readers.

Apart from the editorial board, the designing team has also invested a significant amount of their time in understanding the subject and creating the most relevant covers. They scrutinized every image to scout for the most suitable representation of the subject and create an appropriate cover for the book.

The publishing team has been an ardent support to the editorial, designing and production team. Their endless efforts to recruit the best for this project, has resulted in the accomplishment of this book. They are a veteran in the field of academics and their pool of knowledge is as vast as their experience in printing. Their expertise and guidance has proved useful at every step. Their uncompromising quality standards have made this book an exceptional effort. Their encouragement from time to time has been an inspiration for everyone.

The publisher and the editorial board hope that this book will prove to be a valuable piece of knowledge for researchers, students, practitioners and scholars across the globe.

LIST OF CONTRIBUTORS

T. Gzogyan and S. Gzogyan
Belgorod National Research University, Belgorod, Russia

T. Chikisheva and S. Prokopyev
LCC PC «Spirit», Irkutsk, Russia
Institute of the Earth Crust SB RAS, Irkutsk, Russia
Irkutsk State University, Irkutsk, Russia

E. Kolesov and V. Kilin
PJSC «RUSOLOVO», Moscow, Russia

A. Karpova and V. Tukuser
LCC PC «Spirit», Irkutsk, Russia
Irkutsk State University, Irkutsk, Russia

E. Prokopyev
LCC PC «Spirit», Irkutsk, Russia
Institute of the Earth Crust SB RAS, Irkutsk, Russia

R. Koneev, A. Krivosheeva and A. Sigida
National University of Uzbekistan, Tashkent, Uzbekistan

R. Khalmatov, O. Tursunkulov and N. Iskandarov
Centre for Advanced Technology, Tashkent, Uzbekistan

L. Vaisberg
«Mekhanobr-Tekhnika» REC, St. Petersburg, Russia

E. Kameneva
Petrozavodsk State University, Petrozavodsk, Russia

Y. Pystina and A. Pystin
Institute of Geology Komi SC UB RAS, Syktyvkar, Russia

V. Klimenko, V. Pavlenko and T. Klimenko
Belgorod State Technological University named after V G Shukhov, Belgorod, Russia

A. Gerasimov, V. Arsentyev and V. Lazareva
REC Mekhanobr-tekhnica, St-Petersburg, Russia

İ. Akpınar
Department of Geological Engineering, Faculty of Engineering and Natural Science, Gumushane University, Gumushane, Turkey

V. Chanturiya and T. Matveeva
Institute of Comprehensive Exploitation of Mineral Resources, Russian Academy of Sciences, Moscow, Russia

O. Kotova
Institute of Geology Komi SC UB RAS, Syktyvkar, Russia

E. Podolian and I. Shelukhina
"RMRL" Ltd. (Raw Materials Researching Laboratory), Saint Petersburg, Russia
Department of Mineral Deposits, Saint Petersburg State University, Saint Petersburg, Russia

I. Kotova
Department of Mineral Deposits, Saint Petersburg State University, Saint Petersburg, Russia

A. Sazonov, S. Silyanov and E. Zvyagina
Siberian Federal University, Krasnoyarsk, Russia

V. Pavlov
Special Design and Technology Bureau "Science", Krasnoyarsk, Russia

O. Yakushina and E. Gorbatova
FSBE "All-Russian Institute on Mineral Raw Materials" (VIMS), Moscow, Russia

Z. Nikiforova
Diamond and Precious Metal Geology Institute SB RAS, Yakutsk, Russia

I. Burtsev, I. Perovskiy and D. Kuzmin
Institute of Geology named after Academician N.P. Yushkin Komi Science Center of the Ural Branch of the Russian Academy of Sciences, Syktyvkar, Russia

I. Anisimov, A. Sagitova and O. Troshina
Technology Research Department, Polymetal Engineering JSC, St-Petersburg, Russia

G. Abarzúa and U. Kelm
Instituto de Geología Económica Aplicada (GEA), Universidad de Concepción, Concepción, Chile

L. Gutiérrez
Departamento de Ingeniería Metalúrgica, Facultad de Ingeniería, Universidad de Concepción, Concepción, Chile

J. Morales
Departamento de Geología, Universidad de Salamanca, Salamanca, Spain

K. Diarra and E. Sangu
Department of Geological Engineering, Faculty of Engineering, KOU, 41380 Kocaeli, Turkey

Yu. Voytekhovsky
Saint-Petersburg Mining University, Saint-Petersburg, Russia

H. Tcharo, M. Koulibaly and F. K. N. Tchibozo
Department of Mineral Developing and Oil&Gas Engineering, Engineering Academy, RUDN University, Moscow, Russia

S. Shevchenko, R. Brodskaya, I. Bilskaya, Yu. Kobzeva and V. Lyahnitskaya
Karpinsky Russian Geological Research Institute, St. Petersburg, Russia

I. Razmyslov, O. Kotova and V. Silaev
Institute of Geology Komi SC UB RAS, Syktyvkar, Russia

L. A. Gomze
University of Miskolc, Miskolc, Hungary

T. Yusupov, A. Travin, S. Novikova and D. Yudin
V.S. Sobolev Institute of Geology and Mineralogy SB RAS, Novosibirsk, Russia

A. Sagitova and A. Dolotova
Polymetal Engineering JSC, Saint-Petersburg, Russia

S. Gzogyan and T. Gzogyan
Belgorod National Research University, Belgorod, Russia

I. Ustinov
REC Mekhanobr-Tekhnica, St. Petersburg, Russia

O. Kononov
Moscow State University, Moscow, Russia

F. Javid and E. Çiftçi
Department of Geological Engineering, Faculty of Mines, ITU, 33469 Maslak, Istanbul, Turkey

K. Berkh, D. Rammlmair, M. Drobe and J. Meima
Federal Institute for Geosciences and Natural Resources, Hanover, Germany

A. Khatkova
Department of Mineral Technology, School of Geology, Transbaikal State University, Chita, Russia

L. Nikitina and S. Pateyuk
Department of Geology, University of Kazan, Kazan, Russia

A. Elbendari, V. Potemkin, T. Aleksandrova and N. Nikolaeva
Mineral Processing Department, Saint Petersburg Mining University, Saint Petersburg, Russia

A. Pavlenko and R. Yastrebinskiy
Belgorod State Technological University named after V.G. Shukhov, Belgorod, Russia

A. Dolotova, M. Kharitonova, B. Milman and I. Agapov
Science and Technology Research Division, Polymetal Engineering, Saint-Petersburg, Russia

I. Ustinov
REC «Mekhanobr-Tekhnika», St-Petersburg, Russia

E. Kotova
Mining Museum, St-Petersburg Mining University, St-Petersburg, Russia

E. Ozhogina, A. Rogozhin, O. Yakushina, Yu. Astakhova, E. Likhnikevich, N. Sycheva, A. Iospa and V. Zhukova
FSBE "All-Russian Institute of Mineral Raw Materials" (VIMS), Moscow, Russia

A. Askhabov
Institute of Geology Komi SC UB RAS, Syktyvkar, Russia

L. Andreicheva
Institute of Geology KomiSC UB RAS, Syktyvkar, Russia

V. Maslennikov, S. Maslennikova, N. Aupova, A. Tseluyko and U. Yatimov
Institute of Mineralogy, Ural Branch of RAS, Miass, Russia

R. Large and L. Danyushevsky
CODES ARC Centre of Excellence in Ore Deposits, University of Tasmania, Hobart, Australia

I. Vdovina
Nizhny Novgorod Institute of Education Development, Nizhny Novgorod, Russia

V. Kovalevski and V. Shchiptsov
Institute of Geology, Karelian Research Centre, Petrozavodsk, Russia

S. Sokolov
FSBE "All-Russian Institute of Mineral Raw Materials" (VIMS), Moscow, Russia

V. Gusev, S. Zhmodik and D. Belyanin
Department of Geology and Geophysics, Novosibirsk State University, Novosibirsk, Russia

G. Nesterenko
Institute of Geology and Mineralogy SB RAS (IGM SB RAS), Novosibirsk, Russia

V. Afanasiev, N. Pokhilenko and A. Eliseev
VS Sobolev Institute of Geology and Mineralogy, Siberian Branch, Russian Academy of Sciences, Novosibirsk, Russia

S. Gromilov
Nikolaev Institute of Inorganic Chemistry, Siberian Branch, Russian Academy of Sciences, Novosibirsk, Russia

S. Ugapieva
Diamond and Precious Metal Geology Institute, Siberian Branch, Russian Academy of Sciences, Yakutsk, (Sakha) Yakutia, Russia

V. Senyut
Joint Institute of Mechanical Engineering of the NAS of Belarus, Minsk, Belarus

N. Vorobyov and A. Shmakova
Institute of Geology of Komi SC UB RAS, Syktyvkar, Russia

A. Minibaev
Mining University, Saint-Petersburg, Russia

Yu. Elkina and A. Bulaev
Faculty of Biology, Lomonosov Moscow State University, Moscow, Russia
Research Center of Biotechnology RAS, Moscow, Russia

E. Melnikova and V. Melamud
Research Center of Biotechnology RAS, Moscow, Russia

V. Ponomarchuk, E. Lazareva, N. Karmanov and A. Piryaev
Institute of Geology and Mineralogy SB RAS, Novosibirsk, Russia

H. Selim
Faculty of Engineering, Department of Jewellery Engineering, ITICU, 34840 Küçükyalı, Istanbul, Turkey

H. Sendir
Faculty of Engineering and Architecture, Department of Geological Engineering, EOGU, 26480 Odunpazarı, Eskişehir, Turkey

O. Kazanov
Moscow Branch, FSUE "All-Russian Scientific-Research Institute of Mineral Resources named after N.M. Fedorovsky", Moscow, Russia

G. Logovskaya and S. Korneev
Institute of Earth Science, Saint-Petersburg State University, Saint-Petersburg, Russia

I. Golubeva, I. Burtsev and A. Ponaryadov
Institute of Geology Komi SC UB RAS, Syktyvkar, Russia

A. Shmakova
Institute of Geology Komi SC UB RAS, Syktyvkar, Russia
Karpinsky Russian Geological Research Institute (VSEGEI), Saint-Petersburg, Russia

N. Timonina
Institute of Geology Komi SC UB RAS, Syktyvkar, Russia

S. Karabaev, N. Olmaskhanov, N. Mirsamiev and J. Mugisho
Department of Mineral Developing and Oil and Gas Engineering, Engineering Academy, RUDN University, Moscow, Russia

K. Vorobyev and A. Gomes
Department of Mineral Developing and Oil & Gas Engineering, Engineering Academy, RUDN University, Moscow, Russia

V. Malyukov and K. Vorobyev
Department of Mineral Developing and Oil & Gas Engineering, Engineering Academy, RUDN University, Moscow, Russia

A. Vorobev
Atyrau University of Oil and Gas, Atyrau, Kazakhstan
Peoples' Friendship University of Russia (RUDN University), Moscow, Russia

E. Shchesnyak
Peoples' Friendship University of Russia (RUDN University), Moscow, Russia

W. Nikonow
Federal Institute for Geosciences and Natural Resources (BGR), Hanover, Germany

Y. Denisova, A. Vikhot, O. Grakova and N. Uljasheva
Institute of Geology of the Komi SC UB of RAS, Syktyvkar, Komi Republic, Russia

A. Smetannikov and D. Onosov
PFIC UB RAS "GI UB RAS", Perm, Russia

P. Matyukhin
Belgorod State Technical University named after V.G. Shukhov, Belgorod, Russia

Index

Printed in the USA
CPSIA information can be obtained
at www.ICGtesting.com
JSHW011400091023
49903JS00004B/33